BIM and Construction Health and Safety

This book aims to conceptualise the implementation of building information modelling (BIM) in the workplace health and safety (WHS) management of construction projects to reduce occupational accidents. The safety performance of the construction industry has always been a concern across the globe, and this devastating reputation has drawn the concern of many nations. The potential functions of BIM can drastically alter the WHS practices of the construction industry. BIM facilitates WHS information exchange and management and supports better collaboration and project planning through virtual visualisation of the construction WHS management process. Despite an increasing interest in BIM, a successful mechanism for employing BIM in construction WHS management is absent. Therefore, this book aims to fill this dearth by presenting a model for the integration of such innovative interventions with the current industry practices in a practical manner through the proper identification of effective areas and evaluation of their impacts on the key criteria of construction projects and organisations. This approach will foster the implementation of BIM in the current state of WHS management in the industry and can potentially reduce occupational accidents on construction sites.

This book is essential reading for researchers and professionals interested in how BIM technology can improve health and safety in construction projects. It is intended for engineers, project managers, construction managers, safety officers and safety managers.

Dr Hamed Golzad is Assistant Professor of Building and Construction Management for the School of Design and Built Environment, Faculty of Arts and Design, University of Canberra (UC), Australia.

Dr Saeed Banihashemi is Associate Professor and Postgraduate Program Director of Building and Construction Information Management in the School of Design and Built Environment, Faculty of Arts and Design, University of Canberra (UC), Australia.

Dr Carol Hon is Senior Lecturer at the School of Architecture & Built Environment, Queensland University of Technology (QUT), Australia.

Prof Robin Drogemuller is Virtual Design and Construction Professor at the School of Architecture & Built Environment, Queensland University of Technology (QUT), Australia.

Spon Research

Publishes a stream of advanced books for built environment researchers and professionals from one of the world's leading publishers. The ISSN for the Spon Research programme is ISSN 1940–7653 and the ISSN for the Spon Research E-book programme is ISSN 1940–8005

Corruption in Infrastructure Procurement
Emmanuel Kingsford Owusu and Albert P. C. Chan

Improving the Performance of Construction Industries for Developing Countries
Programmes, Initiatives, Achievements and Challenges
Edited by Pantaleo D Rwelamila and Rashid Abdul Aziz

Work Stress Induced Chronic Diseases in Construction
Discoveries Using Data Analytics
Imriyas Kamardeen

Life-Cycle Greenhouse Gas Emissions of Commercial Buildings
An Analysis for Green-Building Implementation Using A Green Star Rating System
Cuong N. N. Tran, Vivian W. Y. Tam and Khoa N. Le

Data-driven BIM for Energy Efficient Building Design
Saeed Banihashemi, Hamed Golizadeh and Farzad Pour Rahimian

Successful Development of Green Building Projects
Tayyab Ahmad

BIM and Construction Health and Safety
Uncovering, Adoption and Implementation
Hamed Golzad, Saeed Banihashemi, Carol Hon and Robin Drogemuller

For more information about this series, please visit: www.routledge.com

BIM and Construction Health and Safety

Uncovering, Adoption and Implementation

Hamed Golzad, Saeed Banihashemi, Carol Hon and Robin Drogemuller

Routledge
Taylor & Francis Group

LONDON AND NEW YORK

First published 2023
by Routledge
4 Park Square, Milton Park, Abingdon, Oxon OX14 4RN

and by Routledge
605 Third Avenue, New York, NY 10158

Routledge is an imprint of the Taylor & Francis Group, an informa business

British Library Cataloguing-in-Publication Data
A catalogue record for this book is available from the British Library

Library of Congress Cataloging-in-Publication Data
Names: Golizadeh, Hamed, author. | Banihashemi, Saeed, 1985- author. | Hon, Carol K. H., author. | Drogemuller, Robin, author.
Title: BIM and construction health and safety : uncovering, adoption and implementation / Hamed Golizadeh, Saeed Banihashemi, Carol Hon and Robin Drogemuller.
Description: Abingdon, Oxon ; New York, NY : Routledge, [2023] | Series: Spon research | Includes bibliographical references and index.
Identifiers: LCCN 2022061852
Subjects: LCSH: Building information modeling. | Construction industry—Safety measures.
Classification: LCC TH438.13 .G65 2023 | DDC 624.028/9—dc23/eng/20230123
LC record available at https://lccn.loc.gov/2022061852

ISBN: 978-1-032-11536-8 (hbk)
ISBN: 978-1-032-12505-3 (pbk)
ISBN: 978-1-003-22485-3 (ebk)

DOI: 10.1201/9781003224853

Typeset in Times New Roman
by Apex CoVantage, LLC

Hamed would like to dedicate this book to Nasrin and Emma, his lovely wife and daughter.

Contents

Acknowledgements

For the help with providing case study materials and images, we are grateful to VolkerWessels Construction Company, especially Mr Niek Rooijackers and Mr Stijn van Schaijk.

Figures

Tables

Abbreviations

2D	Two-dimensional
3D	Three-dimensional
4D	Four-dimensional
AR	Augmented reality
BIM	Building information modelling
BLS	Bureau of Labor Statistics
CAD	Computer-aided design
CDM	Construction design and management
CHAIR	Construction hazard assessment implication review
ConAC	Construction accident causation
CSF	Critical success factor
DART	Days away, restricted work or transfer
ELCI	Employers Liability Compulsory Insurance
EMR	Experience modification rating
FAIFR	First aid injury frequency rate
FIFR	Fatality incidence frequency
FTE	Full-time equivalent
HFIT	Human Factors Investigation Tool
HSE	Health and Safety Executive
LTIFR	Lost time injury frequency rate
MTIR	Medically treated injury
NCIS	National Coronial Information System
NDOR	Notifiable dangerous occurrence rate
NII	Non-injury incident
NIOSH	National Institute for Occupational Safety and Health
NM/H	Near miss/near hit
NMTIR	Nonmedically treated injury
OSHA	Occupational Safety and Health Administration
PLS-SEM	Partial least square structural equation modelling
PPE	Personal protective equipment
PtD	Prevention through design
R&D	Research and development
RFID	Radio frequency identification

RIRs	Recordable injury rate
RTLS	Real-time location system
RTWRs	Return-to-work rate
SNA	Social network analysis
SWA	Safe Work Australia
SWMS	Safe work method statement
TAFE	Technical and future education
VIF	Variance inflation factor
VR	Virtual reality
WCCR	Workers' compensation claim rate
WCPR	Workers' compensation premium rate
WHS	Workplace health and safety

Preface

The record of health and safety practices in the construction industry is unenviable, and the high number of injuries and fatalities around the globe continues to plague the industry. Building information modelling (BIM), an evolving and innovative technology, is capable of recovering this dire reputation. BIM is the process of creating and managing building information throughout its lifecycle. BIM facilitates project information exchange and management and supports better collaboration and project planning. The potential uses of BIM can drastically alter safety practices in the construction industry; however, an efficacious adoption and implementation process for construction firms is unclear. In this book, we therefore conceptualise the adoption process of BIM applications and implementation for construction projects and firms in light of theoretical backgrounds.

This work is particularly intended to unearth new aspects of BIM integration with construction WHS practice, including relevant gaps, challenges and potential outcomes for this pathway, and put forward a research monograph to develop a framework for the implementation of BIM in construction WHS management. This framework provides a model for the understanding of major and critical causes of construction WHS problems, a clear view of how BIM can improve or alter the current WHS management of construction projects to address these problems and, ultimately, a pathway to implement BIM for WHS at the organisational and project levels.

This book takes an exploratory approach to address the adoption process of BIM to improve construction WHS. Although there are recently published standards for BIM and construction WHS, such as BSI: PAS 1192–6 (2018), a comprehensive adoption roadmap is unavailable. This starts with exploring and uncovering problematic areas of construction WHS practice in different countries requiring improvements. Then, it puts forward the evolving BIM technologies that can recover the areas of concern for construction WHS. Further, this book conceptualises the adoption process of BIM at different stages of identification, evaluation, commitment, preparation, use and post-use evaluation. An in-depth case study method is then presented to bring the ideas into a research-led practical context. This will allow for a stronger explanation of the presented adoption process and a detailed analysis of BIM through the voices and experiences of the actors involved in the industry, as well as real-world accident case studies. This book sparks a new focus on construction safety for academics and researchers, and it is of practical value to industrial practitioners. The book is particularly topical for the construction industry in developed countries and represents a prominent contribution towards improving construction WHS from the BIM lens.

1 Building Information Modelling and Workplace Health and Safety

Backgrounds and Concepts

1.1 Introduction and Background

It is hard to find an industry with a more unenviable record than the construction industry when it comes to health and safety practices. High injury and fatality rates continue to plague the industry worldwide. A comprehensive examination of the statistics published by the labour administrations of many countries and international organisations indicates that a significant proportion of occupational fatalities occurs in the construction sector (1). Work-related construction accidents also lead to delays in project progress, increases in expenses and impacts to contractors' reputations (2, 3).

Lingard (1) reported that over 60,000 workers around the world die every year due to occupational accidents on construction sites. According to the (4) of the United States (US), construction workers accounted for 7.3% of the US national workforce in 2020, while they accounted for 23% of the total occupational fatalities that occurred in the US in the same year. In the United Kingdom (UK), the construction sector reported the highest proportion of occupational fatalities in 2017 and 2018, with mortality rates significantly higher than those of the agriculture and manufacturing industries (5). In Australia, in the five years from 2016 to 2020, 155 work-related fatalities were reported in the construction industry, accounting for an average of 31 deaths per year (6). This five-year average (2016–2020) shows that the construction sector stands as the third most hazardous industry in the country (6). According to the International Labour Organization (ILO), construction sites in developing countries are more dangerous than those in industrialised countries (7). As a few examples, the average work-related fatality rate per 100,000 workers in 2015 for Egypt, Turkey and Ukraine was 37.29, 20.80 and 20.30, respectively. It should be noted that this value for developed countries is usually below five (8).

While traditional safety management tools may not be able to achieve further improvements in reducing occupational fatalities, the adoption of building information modelling (BIM) has untapped potential for improving construction WHS (9, 10).

BIM is the process of creating and managing building information throughout its lifecycle. There are many definitions for BIM, and in many ways, the definition depends on the point of view of who looks at it or the outcome that is sought to be

DOI: 10.1201/9781003224853-1

gained from the approach (11, 12). The Construction Project Information Committee in the UK defines BIM as "the digital representation of physical and functional characteristics of a facility creating a shared knowledge resource for information about it forming a reliable basis for decisions during its life cycle, from earliest conception to demolition" (13). Rajendran and Clarke (14) describe BIM as a data-rich, object-oriented and parametric digital model of the facility, from which users can extract and analyse the appropriate type of data to create information that can facilitate the process of decision-making and delivery in the facility.

BIM provides a more collaborative and visualised platform for the project team while offering advantages for saving, interpreting and analysing various types of information embedded in building models (10, 15, 16). BIM has been widely employed to improve time, cost and other key performance indicators in projects, including WHS (17–20). Moreover, the UK recently became the first country in the world to release BIM standards for construction WHS and risk management (21).

BIM-enabled approaches towards construction workers' health and safety have become popular research areas in recent years due to the high frequency and devastating consequences of these accidents (9, 16, 22). Previous studies on BIM-enabled approaches have demonstrated a significant potential for improvements in construction WHS in several major areas. For example, Albert et al. (23) developed a dynamic and interactive augmented virtual environment using BIM as a hub that helps to develop workers' hazard recognition skills through risk-free learning. The Industry Foundation Classes (IFC)'s format of BIM enables automatic/semi-automatic checking across building models to detect WHS risks in both planning and design models. In the manual approach, three-dimensional (3D) BIM provides the project team and the client with a clearer view of the project compared to that offered by traditional paper-based designs to discuss possible hazards and review prevention through design (PtD) issues from many points of view. Bansal (24) outlined that suitable control measures can be chosen and planned properly once the potential hazards are well identified. Four-dimensional (4D) BIM or 3D BIM plus time component provides a virtual representation of the project schedule, helping the client and project team to identify and assess potential safety risks in the construction phase (9, 25). Goedert and Meadati (26), Fruchter, Schrotenboer (27) and Ding, Zhong (28) introduced BIM as a knowledge management system acting as a repository of documentation throughout the lifecycle of the project. These documents can include several safety management requirements, such as the health condition of the workers, safe work method statements, risk assessment documents, safety records for equipment and machinery, records of accidents, etc. In recent years, several studies have examined the practicality of real-time location systems (RTLS) in construction projects, which integrate navigation systems (e.g., global positioning system, radio frequency identification [RFID], Wi-Fi, etc.), sensors and BIM to prevent accidents caused by unsafe proximity of pedestrian workers to construction equipment, machinery or crane towers (29–32).

To this end, BIM applications offer innovative potential for improving construction WHS management and consequently reducing the number of occupational fatalities.

1.2 Constraints in Exploiting the Potential of BIM for WHS Management

While BIM has many potential applications to improve construction WHS management, a review of the last ten years' peer-reviewed publications in the construction management and safety science fields showed that the investigation of approaches towards the adoption of BIM's WHS capacities in the construction industry are, at best, limited. Using BIM for construction WHS management is an innovative approach, and, at the same time, is a challenging process to integrate with the industry's current WHS practices. The construction industry is one of the most resistant industries against the adoption of innovative approaches, as suggested by Hosseini, et al. (2015). The current practices of the industry often employ less-informative, two-dimensional drawings that hinder active WHS design, planning and management, while BIM tackles these issues by employing more visual and information-rich models (25). To successfully exploit BIM's capacity to enhance WHS management in the construction industry, an investigation is required to identify the critical success factors (CSF) for efficient adoption. Existing research themes have primarily been focused on technical aspects of WHS, such as automating the process of design for safety (33, 34), detecting unsafe proximities (31) and offering virtual training (35, 36), rather than the adoption process. The positive benefits of BIM for construction WHS have been recognised internationally. For example, the UK has developed BIM for construction WHS standards and specifications that support its implementation for major government-funded projects. As part of a global implementation effort, the majority of these standards and specifications are now being translated into international standards (BSI: PAS 1192–6, 2018). Identifying the adoption requirements will enable the industry to efficiently incorporate BIM in the construction WHS management process. A BIM adoption model is required to facilitate BIM adoption for WHS management in the construction industry.

There is, at best, limited research embracing theoretical models to devise a BIM adoption framework for WHS, which remains an unexplored territory. Modelling an adoption plan for the construction industry would likely be complex, considering its innovation-resistant nature. Over time, several adoption models have been developed to provide a theoretical foundation for successful adoption by different industries (37–39). Among them, the innovative adoption theory developed by Slaughter (40) is well-accepted by scientists, rooted in management theories (41) and developed specifically for the construction industry context. Slaughter's innovative adoption theory has been employed in several studies to investigate the critical success factors of new technology adoption (37, 42–44). It provides a theoretical basis for project-level and organisational adoption of an innovation and describes critical stages for the successful adoption of an innovation. The main components of this model include identification, evaluation, commitment, preparation, use and post-use evaluation (40).

As suggested by Slaughter (40), the first step for the implementation of an innovation in construction project management or organisation is to identify the area(s) requiring intervention. There is a lack of investigation focused on creating a robust and, at the same time, up-to-date accident causation model within the context of the construction industry. Therefore, unearthing the critical causes of construction accidents will help to prioritise areas that require intervention (45). National accident surveillance reports usually do not provide detailed information for analysing critical WHS areas and designing preventive approaches. These reports are mainly focused on the immediate reason for injury or death, avoiding reflection on what needs to be done to prevent accidents. For example, annual reports identify the causes of accidents as falls from height, electrocution, falls on slippery surfaces, etc., all of which can be immediately identified after the accident. However, a comprehensive investigation would reveal events leading to an accident. Another important fact is that the applicability of using a WHS model from another country might be questionable when determining critical WHS areas because the roots of accidents vary from one country to another. For instance, in an Australian context, Cooke and Lingard (46) used the construction accident causality (ConAC) framework to investigate the causes of work-related construction fatalities between 2000–2009. The number of accident causations found by Cooke and Lingard (46) was relatively low compared to that reported by similar studies conducted in other countries (47). This was mainly due to a lack of coroner investigation reports for most of the cases in that period and the consideration of police reports that lack accident details. Cooke and Lingard (46) focused on identifying the factors leading to accidents from an individual perspective while disregarding their inter-relationships. However, factors in the accident models are inter-related, forming a network of relationships (48) that must be analysed from a network perspective to identify the most critical factors. Studies examining the network of relationships among the causative factors of construction accidents are rare and not frequently conducted due to the complexity of the cases. The available studies are not up to date in uncovering the key causes of construction accidents.

There is a lack of research into a purposeful evaluation of BIM's potential applications that can show which aspects of BIM could improve the critical WHS areas in the construction industry. Poirier, Staub-French (49) highlighted the necessity of clearly understanding the opportunities that innovation would provide for a construction firm or project. Slaughter (40) described the evaluation of an innovation as assessing the key objectives sought from employing it and stated that the primary objective of WHS management in construction projects is to prevent or control the risks leading to accidents. BIM is not a magic bullet that can hit all targets in the domain of construction WHS management. It is therefore important to investigate and evaluate BIM applications for mitigating or preventing the critical WHS areas that exist in the current practice of the construction industry.

The WHS aspect of BIM is an emerging research area and is brand new for construction firms and projects (10); thus, the critical success factors (CSFs) to its implementation are unrecognised. Despite such scant attention being given to the WHS side of BIM, recognising CSFs to integrate it with the current practice of the industry is of utmost importance, as such an approach could reduce the number of occupational injuries and fatalities. Several studies have attempted to

Challenges	Questions	Research Domains
• The high number of occupational injuries and fatalities • The high cost of occupational accidents • The poor level of digitalisation in WHS management	• What are the main causes of construction accidents? • To what extent can BIM address the identified causes? • What are the CSFs to adopt BIM in WHS management?	• BIM applications for WHS • Root cause of construction accidents • Defusion plan for BIM-WHS

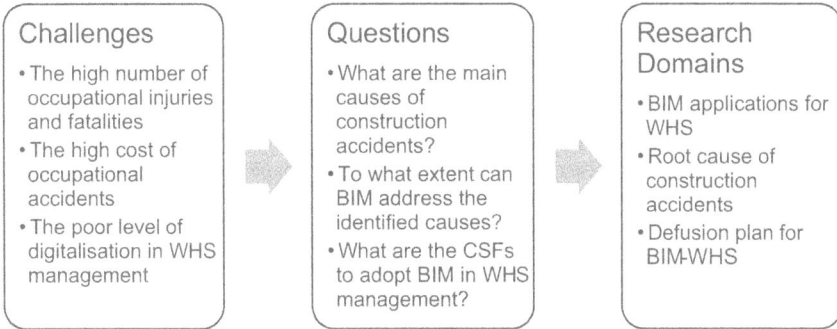

Figure 1.1 Constraints in exploiting the potentials of BIM in WHS management

investigate the diffusion process of BIM, or information and communication technology (ICT), in a broader picture with construction project management practices (49–53). However, CSFs for the implementation of BIM for WHS features within the construction industry context have yet to be determined and require a comprehensive investigation.

In summary, this book identifies three major research gaps within the existing construction project management body of knowledge. First, there is a need to identify the critical causations behind fatal occupational accidents within the context of construction projects. Second, BIM capabilities need to be evaluated against critical accident causations through the lens of industry practitioners to determine the potential advantages of BIM adoption. Third, CSFs for the adaptation of current construction WHS management practices to BIM applications are unrecognised. Examining these three research gaps will bridge the larger gap caused by the lack of a theoretical innovation adoption model to employ BIM in order to reduce occupational fatalities in the construction industry (Figure 1.1). To satisfy the research necessities, three research domains are identified for exploration, including BIM-WHS applications, construction accidents and innovation adoption theories for the construction industry.

1.3 Objectives and Investigation Process

This research book raises important questions that emanate from the problems identified earlier, and these are:

1) What are the key causes of fatal construction accidents?
2) How can BIM support the WHS performance of construction projects?
3) What are the critical barriers and enablers in adopting BIM's WHS applications in construction projects?

Based on these questions, the overarching aim of this research book is to develop a BIM adoption model for reducing occupational fatalities in the construction industry. The research scope is limited to WHS risks identified in the fatal occupational accidents that take place in construction projects. Therefore, the following objectives were identified:

1) To identify the causes of fatal accidents in the construction industry.
2) To evaluate the effectiveness of BIM-enabled approaches and reduce the identified causes of fatal accidents in the construction industry.
3) To determine CSFs for adopting BIM to improve construction WHS performance.
4) To develop an innovation adoption model that integrates BIM's WHS aspects with the current practice of the industry.

Achieving the first objective of this research requires understanding the anatomy of work-related fatal accident causations within the construction industry context. In a previous study on fatal accident causations in the Australian construction industry (46), Cooke and Lingard attempted to extract accident causations from 258 accident cases reported between 2000 and 2009 using content analysis of the reports. Although this research, along with the similar work of Gibb, Lingard (54) and Behm and Schneller (47), made a significant contribution to the existing body of construction safety knowledge in terms of shaping a comprehensive accident model, these models had some major limitations, such as being outdated and lacking a portrayal of the critical causations in the accident network. The current study uses a structured content analysis of recent accident reports to investigate linkages between accident causations, with an aim of addressing the limitation of the previous study. The developed model aims to determine the critical causations of fatal construction accidents and will benefit not only this research but also future investigations of focus areas to improve WHS within the construction industry context.

Considering the unique nature of construction accidents in the contexts of different countries (54), as well as the significant improvements in BIM applications for WHS over the past five years (10), no study has evaluated the efficiency of BIM to control different accident types in the context of the construction industry. The current research develops a conceptual model that illustrates the competency levels of different BIM approaches to mitigate the impact of accident causes. This will provide researchers and practitioners with knowledge regarding the practicality of currently available approaches. The conceptual model aims to facilitate the decision-making processes of construction organisations and projects to determine suitable BIM approaches for potential hazards they may encounter.

As indicated previously, the third objective of the research is to assess the existing knowledge about the factors among and within construction firms at the project level that impede or enable the adoption of BIM for WHS purposes, and to determine further CSFs if possible. While there is a perceived lack of incorporation of innovations in construction projects, revealing CSFs can facilitate innovation adoption and yield significant benefits (55).

Fulfilling these three objectives will lead to the development of a BIM adoption model that improves the WHS performance of the industry. The availability of these tools and guidelines should facilitate their incorporation and allow construction firms to benefit from the innovative approaches that are validated in the research domain despite not being widely practised in the industry (55).

1.4   Implication Areas and Book Structure

This book takes an exploratory approach to uncover, adopt and implement BIM for the role of construction WHS management. It starts with exploring and uncovering the problematic areas of construction WHS practice that require improvements. Then, it puts forward the evolving and innovative BIM mechanisms that can recover these areas of concern. Further, this book conceptualises the adoption process of BIM at different stages of identification, evaluation, commitment, preparation, use and post-use evaluation. An in-depth case study method is then conducted to bring the ideas into a research-led, practical context. This will allow for a greater explanation of the presented adoption process and a detailed analysis of BIM through the voices and experiences of the actors involved in the industry and through real-world accident case studies. This book sparks a new focus on construction safety for academics and researchers and is of practical value to industrial practitioners. This book is designed to satisfy the needs of university students, academics, researchers and industrial practitioners. The contents covered in the proposed new book are comprehensive, practical and timely because the authors have accumulated rich research and consultancy experiences in construction safety. The book is particularly topical for the construction industry in developed countries and represents a prominent contribution towards improving construction WHS from the lens of BIM.

The scope of this research book is limited to construction WHS practice. This book only assesses fatal accidents due to the construction industry's high number of occupational fatalities and the availability of evidence-based case reports. Thus, accidents with minor or major injuries are not included. BIM-enabled approaches used in the study are limited to those published within scientific peer-reviewed journals due to the validation requirements of such journals and conference papers; as such, lecture notes and industrial journal papers were excluded.

The first chapter puts forward an introduction including background for the research book, challenges of construction WHS and BIM-enabled approaches that can potentially alter the current practice of construction WHS. The second chapter sets a theoretical framework for implementing BIM in construction WHS practice. In the third chapter, the construction WHS performance of Australia, the UK and the US is explored through three different cases of national and independent research discoveries that set a clear outline of where the construction industry stands now compared to other industries. Chapter 4 puts forward existing theories of accidents, relationships between factors causing accidents and mitigation approaches specifically related to the construction industry. Chapter 5 concentrates on BIM-enabled approaches to mitigate construction accident causations in different areas of the ConAC model. Chapter 6 focuses on the project and organisation levels for the implementation of BIM in construction WHS by describing the critical roles of involved parties in construction projects, including clients, government, contractors, design teams, procurement methods, technology providers and professional and educational organisations.

The last chapter concludes the book by providing an overview of the presented BIM implementation process for construction WHS.

References

1. Lingard H. Occupational health and safety in the construction industry. *Construction Management and Economics*. 2013;31(6):505–14.
2. Feng Y, Zhang S, Wu P. Factors influencing workplace accident costs of building projects. *Safety Science*. 2015;72:97–104.
3. Wright M, Berman G. *The Case for CDM: Better Safer Design: A Pilot Study*. Suffolk: HSE Books; 2003.
4. Bureau of Labor Statistics (BLS). *National Census of Fatal Occupational Injuries in 2016*; 2020. Available from: https://www.bls.gov/iif/home.htm#2018.
5. Garner-Purkis Z. Deaths in construction rise 27% year on year. *Construction News*; 2018. Available from: www.constructionnews.co.uk/best-practice/health-and-safety/deaths-in-construction-rise-27-year-on-year/10032802.article.
6. Safe Work Australia. *Work-Related Traumatic Injury Fatalities*, Australia; 2020.
7. ILO Estimates Over 1 Million Work-Related Fatalities Each Year [Press Release]; 1999.
8. National Safety Council. *Work-related Deaths Around the World*; 2022. Available from: https://injuryfacts.nsc.org/international/work-related-injuries-around-the-world/work-related-deaths-around-the-world/.
9. Golizadeh H, Hon CKH, Drogemuller R, Hosseini MR. Digital engineering potential in addressing causes of construction accidents. *Automation in Construction*. 2018;95:284–95.
10. Guo H, Yu Y, Skitmore M. Visualization technology-based construction safety management: a review. *Automation in Construction*. 2017;73:135–44.
11. Eastman C, Eastman CM, Teicholz P, Sacks R. *BIM Handbook: A Guide to Building Information Modeling for Owners, Managers, Designers, Engineers and Contractors*. Hoboken, NJ: John Wiley & Sons; 2011.
12. Steel J, Drogemuller R, Toth B. Model interoperability in building information modelling. *Software & Systems Modeling*. 2012;11(1):99–109.
13. Sinclair D. *BIM Overlay to the RIBA Outline Plan of Work*. London: RIBA; 2012.
14. Rajendran S, Clarke B. Building information modeling: safety benefits & opportunities. *Professional Safety*. 2011;56(10):44–51.
15. Malekitabar H, Ardeshir A, Sebt MH, Stouffs R. Construction safety risk drivers: a BIM approach. *Safety Science*. 2016;82:445–55.
16. Zou Y, Kiviniemi A, Jones SW. A review of risk management through BIM and BIM-related technologies. *Safety Science*. 2017;97:88–98.
17. Chong H-Y, Lee C-Y, Wang X. A mixed review of the adoption of building information modelling (BIM) for sustainability. *Journal of Cleaner Production*. 2017;142:4114–26.
18. Eleftheriadis S, Mumovic D, Greening P. Life cycle energy efficiency in building structures: a review of current developments and future outlooks based on BIM capabilities. *Renewable and Sustainable Energy Reviews*. 2017;67:811–25.
19. Johansson M, Roupé M, Bosch-Sijtsema P. Real-time visualization of building information models (BIM). *Automation in Construction*. 2015;54:69–82.

20. Smith P. Project cost management with 5D BIM. *Procedia – Social and Behavioral Sciences*. 2016;226:193–200.
21. BSI: PAS 1192–6. *Specification for Collaborative Sharing and Use of Structured Health and Safety Information Using BIM*. The British Standards Institution (BSI); 2018.
22. Sidani A, Martins JP, Soeiro A. BIM application for construction health and safety: summary for a systematic review. *Journal of Occupational and Environmental Safety and Health IV*; 2023;449:553–63.
23. Albert A, Hallowell MR, Kleiner B, Chen A, Golparvar-Fard M. Enhancing construction hazard recognition with high-fidelity augmented virtuality. *Journal of Construction Engineering and Management*. 2014;140(7):04014024.
24. Bansal VK. Application of geographic information systems in construction safety planning. *International Journal of Project Management*. 2011;29(1):66–77.
25. Ganah A, John GA. Integrating building information modeling and health and safety for onsite construction. *Safety and Health at Work*. 2015;6(1):39–45.
26. Goedert JD, Meadati P. Integrating construction process documentation into building information modeling. *Journal of Construction Engineering and Management*. 2008;134(7):509–16.
27. Fruchter R, Schrotenboer T, Luth GP. From building information model to building knowledge model. *Computing in Civil Engineering*. 2009;2009:380–9.
28. Ding LY, Zhong BT, Wu S, Luo HB. Construction risk knowledge management in BIM using ontology and semantic web technology. *Safety Science*. 2016;87:202–13.
29. Lee G, Cho J, Ham S, Lee T, Lee G, Yun SH, et al. A BIM – and sensor-based tower crane navigation system for blind lifts. *Automation in Construction*. 2012;26:1–10.
30. Li H, Chan G, Huang T, Skitmore M, Tao TYE, Luo E, et al. Chirp-spread-spectrum-based real time location system for construction safety management: a case study. *Automation in Construction*. 2015;55:58–65.
31. Teizer J, Cheng T, Fang Y. Location tracking and data visualization technology to advance construction ironworkers' education and training in safety and productivity. *Automation in Construction*. 2013;35:53–68.
32. Zhang SJ, Teizer J, Pradhananga N, Eastman CM. Workforce location tracking to model, visualize and analyze workspace requirements in building information models for construction safety planning. *Automation in Construction*. 2015;60:74–86.
33. Zhang S, Sulankivi K, Kiviniemi M, Romo I, Eastman CM, Teizer J. BIM-based fall hazard identification and prevention in construction safety planning. *Safety Science*. 2015;72:31–45.
34. Zulkifli AR, Ibrahim CKIC, Belayutham SJ. The integration of building information modelling (BIM) and prevention through design (PtD) towards safety in construction: a review. *Advances in Civil Engineering Materials* 2021:271–83.
35. Bosché F, Abdel-Wahab M, Carozza L. Towards a mixed reality system for construction trade training. *Journal of Computing in Civil Engineering*. 2015;30(2):04015016.
36. Kanade SG, Duffy VG, editors. Use of virtual reality for safety training: a systematic review. *International Conference on Digital Human Modeling and Applications in Health, Safety, Ergonomics and Risk Management*. Springer; 2022.
37. Blayse AM, Manley K. Key influences on construction innovation. *Construction Innovation*. 2004;4(3):143–54.
38. Cao D, Li H, Wang G. Impacts of isomorphic pressures on BIM adoption in construction projects. *Journal of Construction Engineering and Management*. 2014;140(12):04014056.

39. McCoy AP, Badinelli R, Theodore Koebel C, Thabet W. Concurrent commercialization and new-product adoption for construction products. *European Journal of Innovation Management*. 2010;13(2):222–43.

40. Slaughter ES. Implementation of construction innovations. *Building Research & Information*. 2000;28(1):2–17.

41. Damanpour F, Evan WM. Organizational innovation and performance: the problem of "organizational lag". *Administrative Science Quarterly*. 1984:392–409.

42. Banihashemi S, Hosseini MR, Golizadeh H, Sankaran S. Critical success factors (CSFs) for integration of sustainability into construction project management practices in developing countries. *International Journal of Project Management*. 2017;35(6):1103–19.

43. Egbu CO. Managing knowledge and intellectual capital for improved organizational innovations in the construction industry: an examination of critical success factors. *Engineering, Construction and Architectural Management*. 2004;11(5):301–15.

44. Rigby ET, McCoy AP, Garvin MJ. Toward aligning academic and industry understanding of innovation in the construction industry. *International Journal of Construction Education and Research*. 2012;8(4):243–59.

45. Swuste P. "You will only see it, if you understand it" or occupational risk prevention from a management perspective. *Human Factors and Ergonomics in Manufacturing & Service Industries*. 2008;18(4):438–53.

46. Cooke T, Lingard H, editors. A retrospective analysis of work-related deaths in the Australian construction industry. *ARCOM Twenty-seventh Annual Conference*. Association of Researchers in Construction Management (ARCOM); 2011.

47. Behm M, Schneller A. Application of the loughborough construction accident causation model: a framework for organizational learning. *Construction Management and Economics*. 2013;31(6):580–95.

48. Gibb AG, Haslam R, Gyi DE, Hide S, Duff R. What causes accidents? *Proceedings of the Institution of Civil Engineers*. 2006;159(6):46–50.

49. Poirier E, Staub-French S, Forgues D. Embedded contexts of innovation: BIM adoption and implementation for a specialty contracting SME. *Construction Innovation*. 2015;15(1):42–65.

50. Brewer G, Gajendran T, Chen S. The use of ICT in the construction industry: critical success factors and strategic relationship in temporary project organizations. *CIBW78 Information Communication Technology in Construction*. 2005:19–21.

51. Hosseini M, Banihashemi S, Chileshe N, Namzadi MO, Udaeja C, Rameezdeen R, et al. BIM adoption within Australian small and medium-sized enterprises (SMEs): an innovation diffusion model. *Construction Economics and Building*. 2016;16(3):71.

52. Peansupap V, Walker D. Exploratory factors influencing information and communication technology diffusion and adoption within Australian construction organizations: a micro analysis. *Construction Innovation*. 2005;5(3):135–57.

53. Murphy M. Implementing innovation: a stakeholder competency-based approach for BIM. *Construction Innovation*. 2014;14(4):433–52.

54. Gibb A, Lingard H, Behm M, Cooke T. Construction accident causality: learning from different countries and differing consequences. *Construction Management and Economics*. 2014;32(5):446–59.

55. Gambatese JA, Hallowell M. Enabling and measuring innovation in the construction industry. *Construction Management and Economics*. 2011;29(6):553–67.

2 Adoption of BIM for Construction WHS – Theoretical Framework

2.1 BIM as an Innovation

2.1.1 Introduction

"Innovation" is defined by Mahajan and Peterson (1985) as "any idea, object, or practice that is perceived as new by members of the social system" (p. 8). Tornatzky, Fleischer (1) noted that the literature on innovation describes two major processes: Diffusion and implementation. The interference between these two processes is the choice to adopt innovation as a result of organisational perception regarding lagging behind their competitors (2). Wolfe (1994) noted that "the diffusion of an innovation refers to its spread through a population of potential adopters" (p. 407). Wolfe (3) suggested that there are three main research topics regarding innovation: Diffusion of innovations, determination of a company's innovativeness and implementation procedures. The current study deals with the implementation of BIM innovations for improving construction WHS management.

In the construction context, Slaughter (4) describes innovation as the use of a nontrivial alteration, in terms of enhancement to a system or working procedure, that is new to the corresponding organisation. This description is in line with the generic definition given by Damanpour and Evan (5). The former definition is broadly acknowledged as the customised definition of innovation at both the organisational and project level in the context of the construction industry (6–9).

Building upon the taxonomies presented by Marquis (10), Slaughter (11) describes innovations in the construction industry based on five classifications: Incremental, architectural, modular, system and radical innovations.

According to Marquis (10), *incremental innovation* is a small modification to an existing practice that has a small effect on other systems and components. An *architectural innovation* is considered a minimal enhancement of a specific core concept or area that requires substantial amendments to other systems and components to be functional (12). *Modular innovation* describes a substantial enhancement (or even a novel concept) within a specific area that requires no alteration of other systems or components (12). A *system innovation* is a set of corresponding innovations that work as one to offer new functions or attributes and can substantially improve the state of practice or knowledge (4, 13). Lastly, a *radical innovation* is an entirely

DOI: 10.1201/9781003224853-2

new approach or concept that often renders prior resolutions obsolete, including co-dependent systems or components (14).

These innovation classifications can be used to develop the required special expertise, skills and activities to suit the proposed innovation. Afuah and Bahram (15) noted that providing examples from a contractor's point of view can illustrate the concepts of innovation classifications, even though the other members of the project may have a different perception of the degree of change. As an example, a modification in the cutting pattern of a reinforcing steel bar that can provide a modest improvement in the performance of the contractor but does not provide a significant change in the core concept can be considered an incremental innovation. However, the use of fibre-reinforced plastic bars with the conventional profile for cast-in-place concrete elements can be considered a modular innovation that provides significant improvement in the core concept but no alteration in linkage with other systems. The use of new forms of self-compacting concrete that require no vibration or consolidation when pouring concrete into formwork is an architectural innovation that uses similar materials with some modifications but causes a substantial alteration in the process. A combination of the architectural and modular innovation examples mentioned earlier that leads to the construction of a super structure could be considered a system innovation because it combines a complementary set of innovations to provide a new performance level. An example of a radical innovation is to use a prefabricated nano-technology bridge that alters the core concept and linkages to other systems with the capacity to render traditional technology obsolete.

In this book, BIM for construction WHS management is presented as an innovation towards higher standards of WHS practice in the industry. However, each level of BIM implementation in the industry can bring different complexities in the adoption process. Therefore, this chapter reflects the findings of previous studies regarding the classification of BIM approaches in terms of adoption complexity levels. Depending on the level of BIM adoption and how BIM is used, its applications for construction WHS management could be considered architectural, system or radical innovations. The use of BIM as a 3D model for safety design is an architectural innovation, as BIM models facilitate the process of risk detection and decision making (16). On the other hand, employing BIM as an information-rich database and using automated tools to detect design errors can be a significant improvement in the system and the systemic components, making it a system innovation (17). The use of BIM with artificial intelligence linked to a sensors attached to equipment, materials and workers could automatically detect and resolve unsafe situations, which could be considered a radical innovation (18).

2.1.2   *Innovation Implementation Process*

The effective introduction of an innovation in the construction industry can be planned through a chain of activities and implementation stages (19–21). The six major stages that are frequently addressed in empirical studies and theoretical literature are identification, evaluation, commitment, detailed preparation, actual use and post-use evaluation (Figure 2.1) (11).

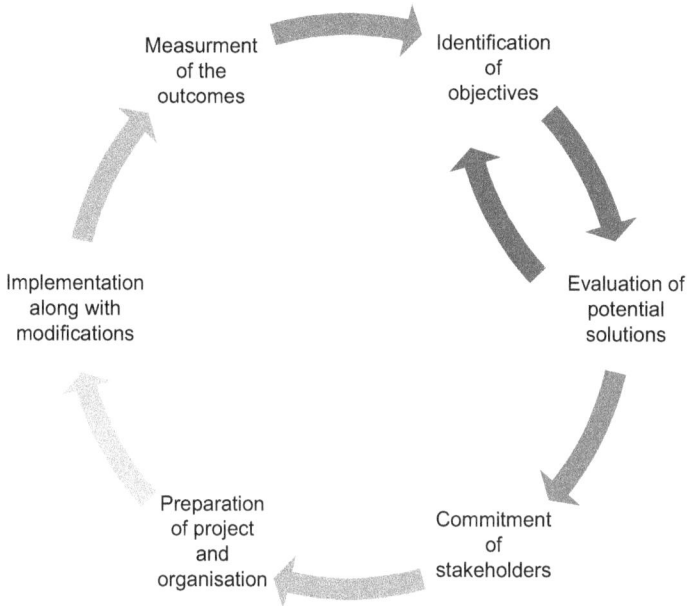

Figure 2.1 Implementation steps for an innovation

2.1.2.1   *Identification Stage*

Slaughter (11) describes the identification stage as clearly identifying in-demand objectives within companies or projects and determining possible substitutes to reach those objectives. In the traditional approach, suppliers and manufacturers are sources for identifying construction-related innovations (22). However, recent studies have shown that general and speciality sub-contractors are the main source of construction innovations (23), specifically in areas that involve interaction and integration among systems, for instance, system and architectural innovations. In addition, the design team, including the architect, structural designers, mechanical designers, etc., can be a major source of innovation, specifically during the early stages of the construction design, to provide a strong solution that satisfies the client's requirements. A critical factor at this stage is the presence of a person in the project or organisation that can identify the areas that require improvement. This person is often called the "gatekeeper" (11). Another critical person required at this stage is the "idea generator", who can provide suitable solutions for the generated challenges and shortcomings (24). These two roles of idea generator and gatekeeper can be performed by the same person in design/construction companies (25, 26).

2.1.2.2   *Evaluation Stage*

The second stage is the evaluation of the innovation to see whether it is capable of improving a system or work procedure (11). Considering the cost-competitiveness of the industry, the traditional major expectation of innovation is to reduce the cost of design and construction activities (27, 28). Nevertheless, recent studies have shown that a large number of the innovations accepted by the industry target improvement in design and construction performance (23). Although companies may consider an innovation for a specific construction project, this can benefit the company by improving its overall performance.

2.1.2.3   Commitment Stage

The third stage of the innovation implementation process is identification of the commitment required by different parties. Tatum (29) describes commitment as the allocation of resources, and this is often achieved through public acknowledgment and announcement of the decision made by a company to implement an innovation. This announcement can often make internal imputes to solve the issue that may occur during the implementation and lead to complete implementation (19). A critical factor at this stage is having a "champion" within the organisation to shepherd the innovation implementation process (24, 25). The resources committed often include finances, equipment, material and personnel resources; long-term provision of such resources is also required to maintain the operation of the innovation. This process can also require the commitment of external organisations, as well as support from the government, to facilitate this process (11).

2.1.2.4   Preparation Stage

A key but often ignored step in the process of implementing innovation is preparing the ground for the innovation. The parties required to be prepared include the people who will apply the innovation in the organisation, the project stakeholders (e.g. client, general contractor, designer, speciality contractor) and external organisations (30). Slaughter (11) argued that two key activities are required at this stage: 1) obtaining the required resources, since construction activities are labour-demanding, and 2) upgrading and training the human resources engaged in this process. At this stage, the project leader is a critical member who will ensure the coordination of the members and negotiate among the stakeholders involved in the project (25).

2.1.2.5   Use Stage

Innovation in the construction industry typically involves modifications in large and complex systems. Therefore, the use stage is when major changes and adjustments take place in the projects to reach the expected objectives (31–33). This requires adjustments in work procedures or systems to achieve the most benefit from employing an innovation and, in some cases, also requires adjustments to the innovation itself to better suit conventional complex systems (21, 34, 35). The critical role during this step is that of the "decision makers" who have authority and influence over the required essential stakeholders and available resources (25). In addition, this stage requires sufficient onsite training of the personnel that will actually use the innovation. Depending on the type of innovation, the level and source of training may vary (30, 36).

2.1.2.6   *Post-Use Evaluation Stage*

Slaughter (11) argued that even though construction projects are temporary in nature and the project team usually disperses after completion of the project, a certain set of information should be kept regarding the implementation of the innovation, including technical, project-specific, organisational, societal and strategic information. Having this information allows for evaluation of the organisation (37, 38). The first factor is a comparison of the original objectives with the outcomes. This measurement of outcomes should be traced back to the identification and evaluation stages. Laborde and Sanvido (39) noted that the involved personnel should be rewarded regardless of their achievements in the implementation process.

2.1.3   *Conceptual Model for the Adoption of BIM for WHS*

BIM has been described (40) as a "radical, transformative and disruptive innovation.". It is, therefore, consistent with Everett Roger's definition of an innovation – "an idea, practice, or object that a person or other unit of adoption perceives as new" – where diffusion is defined as "the process by which an innovation is communicated to a specific group of people and adopted over time by them" (41). Challenges facing the construction industry in recent years, including barriers to sustainable development and occupational fatalities, have led to the examination and adoption of improvements offered by several innovations, such as BIM (23). However, construction projects have not yet achieved an acceptable level of benefit from these innovations (42, 43). This section provides a sound theoretical model on which to base the rest of this study.

BIM-based approaches for improving WHS are not widely practiced within the context of the construction industry; thus, they seem to be new to a large number of organisations (17, 44, 45). Embracing BIM to improve WHS therefore follows the process for the spread of a new phenomenon within the traditional structure of construction firms and can be viewed through the lens of innovation adoption theories (46). In addition, it provides practitioners and investigators with a powerful tool to assess capabilities, predict behaviours, identify effective factors and measure their impact on the process of innovation adoption (47). It would be irrational to conduct a study in the absence of a theoretical background (48).

By reviewing the existing literature, the innovation adoption theory developed by Slaughter (11) can be considered a suitable theoretical departure point for current studies, as it has been acknowledged and employed by several researchers for the implementation of innovations at the organisational and project levels (6, 23, 49, 50). As the current research aims to investigate the process of BIM adoption in the construction industry of Australia to reduce fatalities, it is important to recognise CSFs for the implementation of such approaches. Another advantage of employing the innovation adoption theory is that it provides structural clarity for identifying CSFs, which will facilitate the development of a hierarchy to prioritise the areas requiring critical attention in the adoption of an innovation (51).

Hosseini, Banihashemi (52) and Banihashemi, Hosseini (6) suggested that the last two stages of the original model by Slaughter (11), use and post-use evaluation, could be combined and that the innovation implementation process could be presented in five stages, as presented in Figure 2.2.

In the first stage, it is necessary to identify the critical WHS areas associated with fatal accidents in the Australian construction industry that require improvement through innovation; in the case of this research this focus is on different applications of BIM for construction WHS. This requires identification of the causations behind fatal occupational accidents to identify the critical WHS areas and those within the current WHS practice of the industry that are blameless and can be ignored (53). This is supported by the arguments put forward by Ganah and John (45), indicating that the adoption process of the industry regarding BIM applications requires the identification of these demanded areas.

Subsequent to this identification, the employment of different applications of BIM requires an evaluation of their effectiveness and of the results that could be achieved (54). Factors considered in this study for evaluation are the major accident risk drivers recognised in the identification stage and how effectively different applications of BIM could mitigate and prevent them. Accordingly, potential safety values that may be captured by employing BIM can be assessed.

In the next step, the commitment needed to implement the innovation must be identified; this can also include the allocation of resources by participating parties (11). Previous studies, such as that by Hosseini, Banihashemi (36), have noted that the adoption of BIM at the organisation or project level requires the commitment of not only the contractor but also of other parties, such as the client and design team. This stage can also include public acknowledgment and government support regarding the adoption of an innovation (29).

The preparation step refers to all of the factors that must be prepared. This stage can be presented at two levels: Organisation and project preparation (55). Construction projects are temporary in nature, while organisations require long-term

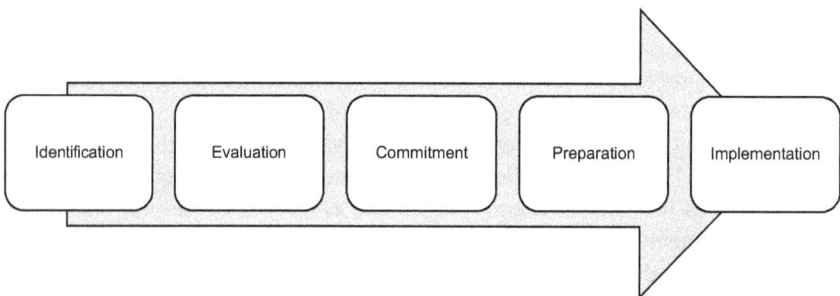

Figure 2.2 Innovation adoption stages

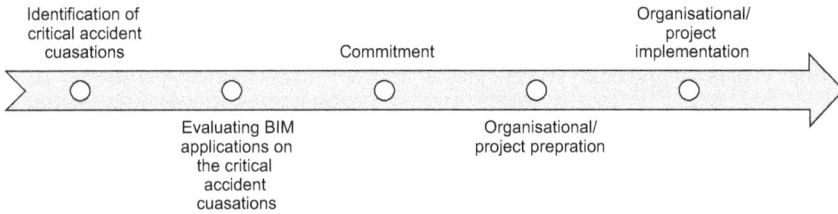

Figure 2.3 Conceptual innovation adoption model of the study

and ongoing preparations that suit their strategic approach in the adoption of an innovation.

Finally, Banihashemi, Hosseini (6) notes the key element of implementation, summarising the use and post-use evaluation stages presented in the original innovation adoption model of Slaughter (11). Proper management of the change process, as well as modification of the procedures to receive maximum benefit from applying the innovation, are the important factors at this stage (21, 34, 35). Similar to the preparation stage, implementation also has two levels – organisation and project implementation.

In summary, the conceptual innovation adoption model for this study follows the sequence of the five steps described earlier (Figure 2.3). Based on the modified innovation adoption model (6), this research postulates that the stages of the conceptual innovation adoption model presented here are related in the manner presented in Figure 2.3. This conceptual model drives the focus of the literature review in the next two sections (Sections 2.3 and 2.4). Focus areas include identifying a suitable accident model, BIM applications for construction WHS, and the critical commitments, preparations and implementations of BIM adoption for construction WHS.

2.2 Identification Stage

In the first stage, it is necessary to identify the critical WHS areas in the construction industry that require improvement through adoption of an innovation, which in the case of this research, is different applications of BIM for construction WHS. This requires identification of the causations behind occupational accidents to determine the critical WHS areas and those within the current WHS practice of the industry that are blameless and can be ignored (53). This is supported by the arguments put forward by Ganah and John (45) indicating that the adoption process for industrial BIM applications requires identification of these demanded areas. Over the years, the construction industry has focused on various types of accident causations, from falling from height, which is still considered a top area for improvement, to mental health issues, which have become a critical research area. Also, government-based organisations, such as the Health and Safety Executive (HSE), the Occupational Safety and Health Administration (OSHA) and others, play a major role in the identification stage. They have first-hand access to information from the accident scenes, as well as further assessments of the case, that enables them to extract the causations of accidents and orient the industry to address them. This stage will be further discussed and analysed in Chapter 4.

2.3   Evaluation Stage

While BIM has advanced significantly over the past decade, there is still a need to identify and evaluate approaches to ensure that the specific applications and capabilities of BIM support the traditional WHS practice of the industry. To this end, the main objective of the evaluation stage is to identify and evaluate the BIM potential for controlling the causes of construction accidents and determine key activities that are required for a prosperous adoption of these applications in the industry. The core purpose of a systematic literature review is to draw upon existing research to deliver a setting for identifying, describing and converting to a higher order of theoretical understanding to deliver the concepts, constructs and associations (56).

In this study, a four-step process was adopted to carry out a systematic literature review, following the approaches taken by Pawson, Greenhalgh (57) and Chong, Lee (58). Subsequent to Step 1, clarifying the aim, Step 2 included a search for evidence wherein search keywords were collated. The keywords for the literature survey were chosen as a result of combining keywords used in three recent review articles by Zou, Kiviniemi (59), Guo, Yu (60) and Golizadeh, Hon (61), consequently creating the following two categories of keywords:

- Building information modelling, digital engineering, BIM, 4D CAD (computer-aided design), 4D BIM, knowledge management, virtual reality (VR), virtual construction, virtual prototyping, augmented reality (AR), model checkers, real-time location system (RTLS)
- Construction accidents, construction safety, construction hazards

After applying the selected keywords, the search was performed in two key academic publication databases: Scopus and Web of Science. The selection of these two sources was due to their reliability and inclusion of most engineering journals (59, 60, 62). The search for journal articles was performed for publications dated from January 2007 to January 2022.

Overall, 403 papers were obtained in the first search. Content analysis was performed on the title and abstract of each article to ensure that articles were relevant to the scope of this study. As a result, 121 articles were found to be relevant to BIM applications related to accident causations, and 44 articles were found to be relevant to success factors for BIM adoption. To ensure that all related articles were included in the data set, Webster and Watson (63) recommended taking into account the reference lists of the collected articles. By taking this approach, 18 additional articles were identified to be relevant, bringing the overall number of articles to 157. After finalising the search for evidence, the articles related to BIM applications were considered for content analysis to identify the potential application for construction WHS.

The frequency of studies on BIM-enabled methods for improving construction WHS have significantly increased during the past years. A literature review by Guo, Yu (60) indicated that since the year 2012, the number of scientific publications on this topic had more than doubled. This study found that BIM-based approaches can be used to support the training process, identify job hazard areas and monitor construction sites (Guo et al., 2017). In another recent holistic review paper, Zou, Kiviniemi (59) examined BIM-based approaches through the lens of the risk management process.

For the current study, a holistic review of the existing literature was conducted on published journal papers published from 2007 to 2022, including the references given within the papers. Unlike Zou, Kiviniemi (59), who considered BIM a technology, and Guo, Yu (60), who examined literature for visualisation aspects of BIM, the review for the current study included relevant information about BIM's implications for WHS practices. The aim was to facilitate the identification of the effectiveness of each application for improving the identified risk factors. As a result of the literature review, these studies were divided into six topics: 3D tools for preliminary risk assessment, automatic/semi-automatic model checkers, 4D (3D+time) construction planning, knowledge management system, AR and VR and RTLS (Figure 2.4). Adoption of BIM for construction WHS is not vastly practiced by the industry and therefore requires a comprehensive approach that will be further investigated and discussed in the Chapter 5.

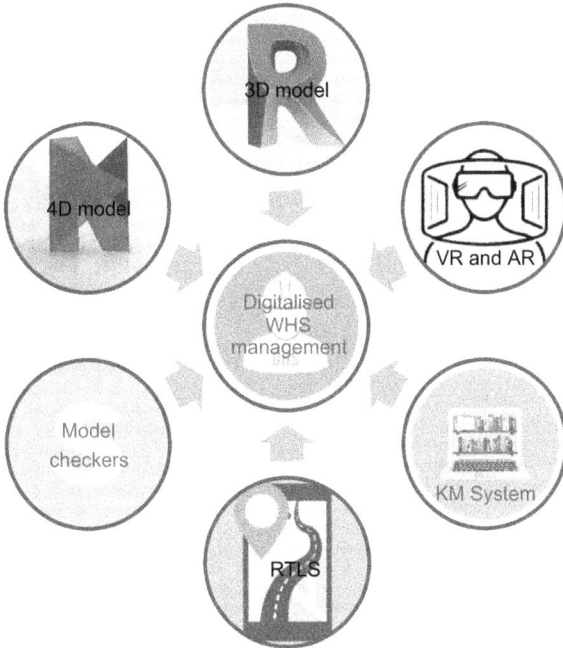

Figure 2.4 BIM and various WHS applications

2.4   Commitment Stage

As one of the biggest clients available in a given country, governments can have a significant influence by mandating the use of BIM for construction WHS management in public projects. Tam, Zeng (64) highlighted the critical role of governments in enforcing rules that companies must improve their WHS performance. Blayse and Manley (7) argued that the process of developing regulations that suit the adoption of new technology is a complex process that mainly depends on the existence of sufficient knowledge among the industry key players and the development of appropriate mechanisms. One of the best examples for such enforcement is the UK's enforcement of the use of BIM in projects (65).

Contractor companies hold most of the liability when a construction accident occurs. They must therefore innovate to improve their WHS performance. Blayse and Manley (2004) highlighted the importance of diffusing new technologies as the main criterion for contractor companies to remain competitive in the industry and present their improvements in operations and distinctive technical capabilities. As identified in this study, the process of the adoption of BIM for WHS management by the industry is highly dependent on the contractors' commitment.

Whether it is called "prevention through design" or "safety in design", the process highlights the large commitment of the design team to devise structures that are safe to construct. Schulte, Rinehart (66) defined this process as:

> The practice of anticipating and designing out potential occupational safety and health hazards and risks associated with new processes, structures, equipment, or tools, and organising work, such that it takes into consideration the construction, maintenance, decommissioning, and disposal/recycling of waste material, and recognising the business and social benefits of doing so.
>
> (p. 115)

In contrast, vendors and technology providers play a significant role in the establishment of BIM applications for construction WHS. The technology here is very new for most construction companies, and this requires the development of more user-friendly platforms. As such, Eastman, Lee (67) and Vakilinezhad, Dias (68) argued that the current tools do not address WHS issues and are more focused on design aspects, as well as the time and cost management of the projects. Tools such as Autodesk Navisworks, Asta PowerProject, Synchro 4D construction project management and 3D BIM software allow for visual detection of WHS concerns, while WHS management processes require the use of embedded information.

2.5   Preparation of Projects and Organisations

A key but often ignored step in the process of implementing an innovation is preparing the ground for the innovation. Under-capitalisation, fragmented organisational structures, and poor communication are among the problems identified in various business areas (69). It appears that most of these problems arise as a result of barriers to organisational preparation, as suggested by Boyed, Robson (70). According to Boyed, Robson (70), the only sustainable competitive advantage an organisation can maintain is its ability to learn faster than its competitors in order to produce world-class construction quality. For the construction industry, the former study have identified the importance of creating a tradition of "learning innovation". The parties required to be prepared include the people who will apply the innovation in the organisation, the project stakeholders (e.g. client, general contractor, designer, speciality contractor) and external organisations (30). Slaughter (11) argued that two key activities are required at this stage: 1) obtaining the required resources, and, because construction activities are labour-demanding 2) upgrading and training the human resources engaged in this process. At this stage, the project leader has a critical role of coordinating of the members and negotiating among the stakeholders involved in the project (25).

Research and development (R&D) of new concepts, technologies and processes is instrumental in the innovation process. However, the ability to perform R&D activities may be difficult for some firms, especially in such a project-based industry as construction. It is often difficult for firms to pursue R&D on current projects given the constraints of the project's objectives and goals. If a firm is interested in innovating, it may find that current projects are obstacles to doing so and may not have the resources or opportunities to conduct the necessary R&D elsewhere. Support from the owner/client for the innovation is often a necessity (46).

Part of what makes up an organisation's culture is the climate (or environment) in which the employees work. Climate is characterised by the employment surroundings, both physical and organisational, within which the employees act. Examples of factors that affect organisational climate with respect to innovation include upper management's emphasis on innovation and whether formal recognition is given to those employees who innovate (71).

Formally including innovation in an organisation's strategic plan and administration emphasises the importance of innovation for the employees, which can motivate workers to be involved in the innovation process (72). However, an organisation's structure should not be overly restrictive, complicated, multi-layered or stifling of opportunities for developing and implementing new ideas. The benefit of having mechanisms that eliminate such barriers and facilitate communication and sharing is supported by Bosch-Sijtsema and Postma (73). In a study of how firms in the construction industry cooperate, Bosch-Sijtsema and Postma (73) found that cooperation, mutual sharing of knowledge and mutual access to knowledge enabled innovation.

2.6   Implementation at the Project and Organisational Levels

Innovation in the construction industry typically relates to modifications in large and complex systems. The extent to which an innovation has diffused throughout an industry can serve as an indicator of its success. The use of a new technology is more likely to cause positive change in the firm's financial success if there is greater diffusion of the technology (74). In addition, the use stage is when major changes and adjustments take place in the projects to reach the expected objectives (31–33). This requires adjustments in work procedures or systems to achieve the most benefit from employing an innovation; in some cases, it also requires adjustments to the innovation itself to better suit conventional complex systems (21, 34, 35). The critical role during this step is that of the "decision makers", who have authority and influence over the required essential stakeholders and available resources (25). Koebel, Papadakis (75) identified management strategies that distinguish successful firms in the home building industry from unsuccessful ones. It was found that successful firms possessed the following characteristics motivated leaders; 2) technology advocates within the company; 3) emphasis on being creative and incorporating new products; 4) access to technology transfer programs; and 5) employment of unions.

This stage requires sufficient onsite training of the personnel that will actually use the innovation. Depending on the type of innovation, the level and source of training may vary (30, 36). Slaughter (11) argued that even though construction projects are temporary in nature and the project team usually disperses after completion of the project, a certain set of information should be kept regarding the implementation of the innovation, including technical, project-related, organisational, societal and strategic information. Having this information allows for evaluation of the organisation (37, 38). The first factor is to compare the original objectives with the outcomes. This measurement of the outcomes should be traced backed to the identification and evaluation stages. Laborde and Sanvido (39) noted that the involved personnel should be rewarded regardless of the achievements in the implementation process. The operant conditioning theory stipulates that behaviour changes are caused by internal events that occur in an individual's environment (76). Responses and consequences are essential elements of operant conditioning. A favourable or positively reinforced consequence will increase the likelihood of a similar response more so than a punishment. As a result, workers' attitudes toward embracing innovation could be improved by understanding the use of positive and negative reinforcements.

2.7 Summary

This chapter was meant to introduce the underlying theories and frameworks of BIM for improving the WHS management of construction companies. A particular focus was placed on assessing the innovation value and complexity of BIM. As part of the innovation adoption process, organisations and projects are required to identify BIM applications for construction WHS practices, evaluate them, identify the commitments needed and prepare and implement requirements at the organisational and project levels. To conclude, it is essential to identify the key steps and barriers within and outside of the adoption process through a detailed analysis.

References

1. Tornatzky LG, Fleischer M, Chakrabarti A. *The Processes of Technological Innovation.* Issues in organization and management series. Lexington, MA: Lexington Books; 1990.

2. Winch G. Zephyrs of creative destruction: understanding the management of innovation in construction. *Building Research & Information*. 1998;26(5):268–79.
3. Wolfe RA. Organizational innovation: review, critique and suggested research directions. *Journal of Management Studies*. 1994;31(3):405–31.
4. Slaughter ES. Models of construction innovation. *Journal of Construction Engineering and Management*. 1998;124(3):226–31.
5. Damanpour F, Evan WM. Organizational innovation and performance: the problem of" organizational lag". *Administrative Science Quarterly*. 1984:392–409.
6. Banihashemi S, Hosseini MR, Golizadeh H, Sankaran S. Critical success factors (CSFs) for integration of sustainability into construction project management practices in developing countries. *International Journal of Project Management*. 2017;35(6):1103–19.
7. Blayse AM, Manley K. Key influences on construction innovation. *Construction Innovation*. 2004;4(3):143–54.
8. Egbu CO. Managing knowledge and intellectual capital for improved organizational innovations in the construction industry: an examination of critical success factors. *Engineering, Construction and Architectural Management*. 2004;11(5):301–15.
9. Rigby ET, McCoy AP, Garvin MJ. Toward aligning academic and industry understanding of innovation in the construction industry. *International Journal of Construction Education and Research*. 2012;8(4):243–59.
10. Marquis D. *Readings in the Management of Innovation*, ed. by ML Tushman and WL Moore. Boston: Ballinger Publishing Company; 1988.
11. Slaughter ES. Implementation of construction innovations. *Building Research & Information*. 2000;28(1):2–17.
12. Henderson R, Clark KB. Architectural innovation: the reconfiguration of existing product technologies and the failure of established firms. *Administrative Science Quarterly*. 1990;35(1):9–30.
13. Cainarca GC, Colombo MG, Mariotti S. An evolutionary pattern of innovation diffusion: the case of flexible automation. *Research Policy*. 1989;18(2):59–86.
14. Nelson R, Winter S. *In Search of a Useful Theory of Innovation Research Policy*. Amsterdam: North-Holland; 1977.
15. Afuah AN, Bahram N. The hypercube of innovation. *Research Policy*. 1995;24(1):51–76.
16. Sacks R, Whyte J, Swissa D, Raviv G, Zhou W, Shapira A. Safety by design: dialogues between designers and builders using virtual reality. *Construction Management and Economics*. 2015;33(1):55–72.
17. Malekitabar H, Ardeshir A, Sebt MH, Stouffs R. Construction safety risk drivers: a BIM approach. *Safety Science*. 2016;82:445–55.
18. Park J, Kim K, Cho YK. Framework of automated construction-safety monitoring using cloud-enabled BIM and BLE mobile tracking sensors. *Journal of Construction Engineering and Management*. 2017;143(2).
19. Goodman PS, Griffith TL. A process approach to the implementation of new technology. *Journal of Engineering and Technology Management*. 1991;8(3–4):261–85.
20. Meyer AD, Goes JB. Organizational assimilation of innovations: a multilevel contextual analysis. *Academy of Management Journal*. 1988;31(4):897–923.
21. Von Hippel E, Tyre MJ. How learning by doing is done: problem identification in novel process equipment. *Research Policy*. 1995;24(1):1–12.
22. Pries F, Janszen F. Innovation in the construction industry: the dominant role of the environment. *Construction Management and Economics*. 1995;13(1):43–51.
23. Hosseini MR, Chileshe N, Zuo J, Baroudi B. Adopting global virtual engineering teams in AEC projects: a qualitative meta-analysis of innovation diffusion studies. *Construction Innovation*. 2015;15(2):151–79.

24. Roberts EB. What we've learned: managing invention and innovation. *Research-Technology Management.* 1988;31(1):11–29.
25. Nam CH, Tatum CB. Leaders and champions for construction innovation. *Construction Management & Economics.* 1997;15(3):259–70.
26. Stewart WS, Tatum C. Segmental placement of Renton outfall: construction innovation. *Journal of Construction Engineering and Management.* 1988;114(3):390–407.
27. Duke R. Local building codes and the use of cost-saving methods. *Report of the Bureau of Economics to the Federal Trade Commission*; 1988. Available from: https://www.ftc.gov/sites/default/files/documents/reports/local-building-codes-and-use-cost-saving-methods/232142.pdf.
28. Seaden G. Economics of innovation in the construction industry. *Journal of Infrastructure Systems.* 1996;2(3):103–7.
29. Tatum CB. Process of innovation in construction firm. *Journal of Construction Engineering and Management.* 1987;113(4):648–63.
30. Cross M. *Technical Change, the Supply of New Skills and Product Diffusion.* London Papers in Regional Science 12 Technological Change and Regional Development; 1983: 54–67.
31. Hutcheson P, Pearson AW, Ball DF. Sources of technical innovation in the network of companies providing chemical process plant and equipment. *Research Policy.* 1996;25(1):25–41.
32. Kangari R, Miyatake Y. Developing and managing innovative construction technologies in Japan. *Journal of Construction Engineering and Management.* 1997;123(1):72–8.
33. Slaughter S. Innovation and learning during implementation: a comparison of user and manufacturer innovations. *Research Policy.* 1993;22(1):81–95.
34. Fleck J. Learning by trying: the implementation of configurational technology. *Research Policy.* 1994;23(6):637–52.
35. Voss CA. Implementation: a key issue in manufacturing technology: the need for a field of study. *Research Policy.* 1988;17(2):55–63.
36. Hosseini M, Banihashemi S, Chileshe N, Namzadi MO, Udaeja C, Rameezdeen R, et al. BIM adoption within Australian small and medium-sized enterprises (SMEs): an innovation diffusion model. *Construction Economics and Building.* 2016;16(3):71.
37. Buijs A, Silvester S. Demonstration projects and sustainable housing: the significance of demonstration projects as a so called second generation steering instrument is examined by the use of case studies. *Building Research and Information.* 1996;24(4):195–202.
38. Rubenstein AH, Chakrabarti AK, O'Keefe RD, Souder WE, Young H. Factors influencing innovation success at the project level. *Research Management.* 1976;19(3):15–20.
39. Laborde M, Sanvido V. Introducing new process technologies into construction companies. *Journal of Construction Engineering and Management.* 1994;120(3):488–508.
40. Gledson B, Greenwood DJ. Surveying the extent and use of 4D BIM in the UK. *JoITiC.* 2016;21:57–71.
41. Rogers EM, Singhal A, Quinlan MM. *Diffusion of Innovations. An Integrated Approach to Communication Theory and Research.* New York: Routledge; 2014: 432–48.
42. Davidson C. Innovation in construction – before the curtain goes up. *Construction Innovation.* 2013;13(4):344–51.
43. Rundquist J, Emmitt S, Halila F, Hjort B, Larsson B. Construction innovation: addressing the project-product gap in the Swedish construction sector. *International Journal of Innovation Science.* 2013;5(1):1–10.
44. Enshassi A, Ayyash A, Choudhry RM. BIM for construction safety improvement in Gaza strip: awareness, applications and barriers. *International Journal of Construction Management.* 2016;16(3):249–65.

45. Ganah A, John GA. Integrating building information modeling and health and safety for onsite construction. *Safety and Health at Work*. 2015;6(1):39–45.
46. Gambatese JA, Hallowell M. Enabling and measuring innovation in the construction industry. *Construction Management and Economics*. 2011;29(6):553–67.
47. Mahajan V, Peterson RA. *Models for Innovation Diffusion*. Newbury Park, CA: Sage Publications; 1985.
48. Anfara Jr VA, Mertz NT. *Theoretical Frameworks in Qualitative Research*. Sage Publications; 2014.
49. Xue X, Zhang X, Wang L, Skitmore M, Wang Q. Analyzing collaborative relationships among industrialized construction technology innovation organizations: a combined SNA and SEM approach. *Journal of Cleaner Production*. 2018;173:265–77.
50. Lindgren J, Emmitt S, Widén K. Construction projects as mechanisms for knowledge integration: mechanisms and effects when diffusing a systemic innovation. *Engineering, Construction and Architectural Management*. 2018;25(11):1516–33.
51. Liu H, Skibniewski MJ, Wang M. Identification and hierarchical structure of critical success factors for innovation in construction projects: Chinese perspective. *Journal of Civil Engineering and Management*. 2016;22(3):401–16.
52. Hosseini MR, Banihashemi S, Martek I, Golizadeh H, Ghodoosi F. Sustainable delivery of megaprojects in Iran: integrated model of contextual factors. *Journal of Management in Engineering*. 2017;34(2):05017011.
53. Swuste P. "You will only see it, if you understand it" or occupational risk prevention from a management perspective. *Human Factors and Ergonomics in Manufacturing & Service Industries*. 2008;18(4):438–53.
54. Bhatt N, Ved A. ICT in new product development: revulsion to revolution. *Driving the Economy through Innovation and Entrepreneurship*. India, Connaught Place, Delhi: Springer; 2013: 833–45.
55. Sawhney A, Mukherjee KK, Rahimian FP, Goulding JS. Scenario thinking approach for leveraging ICT to support SMEs in the Indian construction industry. *Procedia Engineering*. 2014;85:446–53.
56. Petticrew M, Roberts H. *Why Do We Need Systematic Reviews?* Oxford: Blackwell Publishing Ltd; 2008: 1–26 p.
57. Pawson R, Greenhalgh T, Harvey G, Walshe K. Realist review – a new method of systematic review designed for complex policy interventions. *Journal of Health Services Research & Policy*. 2005;10(1_suppl):21–34.
58. Chong H-Y, Lee C-Y, Wang X. A mixed review of the adoption of building information modelling (BIM) for sustainability. *Journal of Cleaner Production*. 2017;142:4114–26.
59. Zou Y, Kiviniemi A, Jones SW. A review of risk management through BIM and BIM-related technologies. *Safety Science*. 2017;97:88–98.
60. Guo H, Yu Y, Skitmore M. Visualization technology-based construction safety management: a review. *Automation in Construction*. 2017;73:135–44.
61. Golizadeh H, Hon CKH, Drogemuller R, Hosseini MR. Digital engineering potential in addressing causes of construction accidents. *Automation in Construction*. 2018;95:284–95.
62. Skibniewski MJ. Information technology applications in construction safety assurance. *Journal of Civil Engineering and Management*. 2014;20(6):778–94.
63. Webster J, Watson RT. Analyzing the past to prepare for the future: writing a literature review. *MIS Quarterly*. 2002;26(2):xiii–xxiii.
64. Tam CM, Zeng SX, Deng ZM. Identifying elements of poor construction safety management in China. *Safety Science*. 2004;42(7):569–86.

65. Ganah A, John GA. Achieving level 2 BIM by 2016 in the UK. *Computing in Civil and Building Engineering*. 2014;2014:143–50.
66. Schulte PA, Rinehart R, Okun A, Geraci CL, Heidel DS. National prevention through design (PtD) initiative. *Journal of Safety Research*. 2008;39(2):115–21.
67. Eastman C, Lee J-M, Jeong Y-S, Lee J-K. Automatic rule-based checking of building designs. *Automation in Construction*. 2009;18(8):1011–33.
68. Vakilinezhad M, Dias P, Ergan S, editors. Achieving model-based safety at construction sites: BIM and safety requirements representation. *Proc of the 33rd CIB W78 Conference*. Brisbane, Australia; 2016.
69. Vakola M, Rezgui YJTLO. Organisational learning and innovation in the construction industry. *The Learning Organization*. 2000;7(4):174–84.
70. Boyed D, Robson AJD, E, Spon F, et al. Enhancing learning in construction projects. *The Organization and Management of Construction: Shaping theory and practice* (vol. 1), ed. by D Langford and A Retik. London: Routledge; 1994.
71. Soo-Hoon L, Poh-Kam W, Chee-Leong C. Human and social capital explanations for R&D outcomes. *IEEE Transactions on Engineering Management*. 2005;52(1):59–68.
72. Steele J, Murray M, editors. The application of structured exploration to develop a culture of innovation. *National Conference, Chartered Institute of Building Services Engineers*. Citeseer; 2001.
73. Bosch-Sijtsema PM, Postma TJBM. Cooperative innovation projects: capabilities and governance mechanisms*. *Journal of Product Innovation Management*. 2009;26(1):58–70.
74. Gambatese JA, Hallowell M. Factors that influence the development and diffusion of technical innovations in the construction industry. *Construction Management and Economics*. 2011;29(5):507–17.
75. Koebel CT, Papadakis M, Hudson E, Cavell M. The diffusion of innovation in the residential building industry. *Virginia Center for Housing Research*. 2003;41–9. https://www.vchr.vt.edu/sites/vchr/files/upload/publications/Diffusion%20Report.pdf.
76. O'Donohue W, Ferguson KE. *The Psychology of BF Skinner*. Newbury Park, California: Sage Publications; 2001.

3 The Current State of Practice in Construction WHS of Australia, the UK and the US

3.1 WHS Performance Indicators in the Construction Industry

According to Chan and Hon (1), measurement indicators are essential for assessing safety performance, and can be classified as either lagging or leading indicators. The Australian Constructors Association (2) defines lagging indicators as events that have already occurred, have harmed employees of the organisation, and can be used to assess safety performance. In contrast, leading indicators are defined as proactive measures undertaken to improve their safety outcomes.

National accident surveillance reports usually provide lagging indicators such as lost time injury rates or fatality rates. First aid injury frequency rates, fatality incidence frequencies and lost time injury frequency rates are among the most common lagging indicators for assessing safety performance in Australia. Medically treated injuries, nonmedically treated injuries, notifiable dangerous occurrence rates, non-injury incidents or near miss/near hits, return to work rates, workers' compensation claim rates and workers' compensation premium rates are also among the most common lagging indicators for evaluating safety performance (3).

Safety performance in the United States is primarily assessed based on metrics such as recordable injury rates reported by OSHA, days away, restricted work, or transfer injury rates and experience modification ratings (4).

A disadvantage of using lagging indicators is that they reflect the past, are reactive rather than proactive and are not predictable (2). Lagging indicators are frequently viewed as nominal because measures are only taken when they indicate that performance is below expectations. Aside from this, lagging indicators are unable to provide any insight into the reasons for poor performance (1).

Leading indicators monitor the performance of the safety processes, unlike lagging indicators (4). As a result, they are more useful for identifying safety problems and determining the appropriate courses of action. In order to improve safety, leading indicators aid in prioritising efforts. It is important to note that there are two types of leading indicators: passive and active. An organisation's passive indicators convey information about its safety performance on a macro level, but they are not short-term and cannot be modified rapidly. On the other hand, active leading indicators can be changed rapidly and easily. Table 3.1 present some examples of active and passive indicators.

DOI: 10.1201/9781003224853-3

Table 3.1 Examples of passive and active leading indicators

	Passive leading indicator	Active leading indicator
Ex. 1	Number or proportion of the managerial staff certified for 10 hours (or 30 hours) of OSHA safety training	Number or proportion of site tool-box meetings that site supervisors/project managers join
Ex. 2	Number or proportion of front-line staff certified for 10 hours (or 30 hours) of OSHA safety training	Number or proportion of site pre-task planning meetings that site supervisors/project managers join
Ex. 3	Number or proportion of sub-contractors selected using safety as a selection criterion	Number or proportion of safety compliance on safety inspections

Source: Hinze, Thurman (4)

Although leading indicators are a more progressive way to compare the safety performance of different industries, on a national scale, such data is not available for any country. Thus, the current research will use the lagging indicators to assess the current status of the construction industry's safety performance as compared to that of other industries.

3.2   WHS Performance of the Australian Construction Industry Compared to That of Other Industries

Safe Work Australia (5) defines the construction industry as work involved in the construction, alteration or demolition of buildings and other structures or the preparation of building sites. Moreover, a serious claim is identified as a workers' compensation claim for an incapacity resulting in a total absence from work of one week or more. In 2002, a national WHS strategy was initiated with the support of the Australian Chamber of Commerce and Industry, the Australian Council of Trade Unions and the Workplace Relations Ministers' Council with the aim of developing a plan for improving WHS performance in a large number of occupations (6). According to outcomes from the 2002–2012 strategy, milestones to be achieved by the year 2022 were set as the realisation of

- at least a 20% decrease in the rate of worker fatalities;
- at least a 20% decrease in the number of serious claims for incidents that cause one or more weeks off work; and
- at least a 20% decrease in the number of claims for musculoskeletal disorders causing one or more weeks off work (6).

According to the strategic framework, the construction industry's WHS status is recognised as hazardous by nature; thus, there are demands for nationwide action towards preventive interventions. However, despite all efforts to set new targets, safety performance improvement has plateaued (7). The construction industry stood in second place for the relative number of serious claims in 2019–2020 with 13% of all claims, following the health care industry (18% of all claims) and followed by the social assistance industry (12% of all claims). Despite making up only 30% of the workforce, these three industries accounted for 43% of all serious claims.

According to Safe Work Australia statistics, the number of workers in the construction industry grew by 33% over the 11 years prior to 2015. Between 2016 and 2020, 155 work-related deaths were reported in Australia's construction industry, representing an average of 31 deaths per year (7). Based on this data, the construction industry ranks third among all industries in terms of industry hazards (7). The majority of these incidents (58%) occurred in the construction services industry sub-division (Table 3.2). Younger workers aged under 25 accounted for 13% of fatalities in the construction industry, compared with only 8% of fatalities across all industries during the five-year period (Table 3.3).

Table 3.2 Worker fatalities: Construction industry sub-divisions and groups, 2016 to 2020 (combined total)

Industry sub-divisions and groups	No. of fatalities	% of fatalities
Construction Services	**89**	**58%**
Building installation services	23	15%
Other construction services	19	12%
Building structure services	19	12%
Land development & site preparation services	16	10%
Building completion services	12	8%
Building Construction Total	**46**	**30%**
Residential building construction	26	17%
Non-residential building construction	20	13%
Heavy & Civil Engineering Construction Total	**19**	**12%**
Construction five-year total	154	100%

Source: (7)

Table 3.3 Worker fatalities: Construction industry and all industries by age group, 2016 to 2020 (combined total)

Age group	Construction industry – No. of fatalities	Construction industry – % of fatalities	All industries – % of fatalities
Under 25	20	13%	8%
25–34	22	14%	16%
35–44	17	11%	17%
45–54	35	23%	19%
55–64	45	29%	25%
65 & over	15	10%	14%
Five-year total	**154**	**100%**	**100%**

Source: Safe Work Australia (7)

Table 3.4 Worker fatalities: Construction industry sub-divisions by the mechanism of incidents, 2016 to 2020 (combined total)

Construction sub-division and mechanism	No. of fatalities	% of fatalities
Construction services	**89**	**58%**
Fall from a height	28	18%
Vehicle collision	16	10%
Contact with electricity	12	8%
Being hit by falling objects	12	8%
Being hit by moving objects	9	6%
Being trapped between stationary and moving objects	4	3%
Slide-in or cave-in	2	1%
Being trapped by moving machinery	2	1%
Other mechanisms	4	3%
Building construction	**46**	**30%**
Fall from a height	19	12%
Being hit by falling objects	10	6%
Vehicle collision	4	3%
Being hit by moving objects	4	3%
Contact with electricity	4	3%
Being trapped by moving machinery	2	1%
Other mechanisms	3	2%
Heavy & civil engineering construction	**19**	**12%**
Being hit by moving objects	9	6%
Vehicle collision	4	3%
Slide-in or cave-in	2	1%
Being hit by falling objects	2	1%
Other mechanisms	2	1%

Source: Safe Work Australia (7)

Between 2016 and 2020, falling from a height was the main cause of fatality in both the construction services and building construction industry sub-divisions, resulting in 48 deaths across the construction industry (Table 3.4).

3.3   WHS Performance of the US Construction Industry Compared to That of Other Industries

The Bureau of Labor Statistics (BLS) (8) defines the construction sector as "establishments primarily engaged in the construction of buildings or engineering projects (e.g., highways and utility systems)". Establishments primarily engaged in the preparation of sites for new construction and in the sub-division of land for sale as building sites also are included in this sector.

In 1970, the Occupational Safety and Health Act established both the National Institute for Occupational Safety and Health (NIOSH) and OSHA. By conducting and translating research, as well as providing information, education and training in the field of occupational safety and health, NIOSH makes sure that working conditions are safe and healthy for both men and women.

According to the BLS (9), the construction industry had the most fatalities among private industries in 2019, accounting for 20.6% of total occupational deaths. The rate of fatal injuries per 100,000 full-time equivalents (FTEs) during this period was 9.67. According to Brown, Harris (10), the construction industry ranks first among all industries in terms of industry hazards. The four-year average of worker fatalities for 2016–2019 shows that the majority of these (76.2%) occurred among construction trade workers (Table 3.5). The rate of fatal injuries per 100,000 FTEs for older workers aged over 65 for this period was 22, compared with only 7.89 across all industries during the four-year period (Table 3.6).

In terms of fatalities, between 2016 and 2019, falling to a lower level was the primary mechanism of fatality in construction works, resulting in 1461 deaths across the industry (Table 3.7).

Table 3.5 Worker fatalities in the US: Construction industry occupation groups, 2016 to 2020

Occupation	No. of fatalities	% of fatalities
Supervisors of construction and extraction workers	535	13.2%
Construction trades workers	2,969	73.1%
Extraction workers	196	13.7%

Source: Bureau of Labor Statistics (9)

Table 3.6 Average worker fatalities in the US: Construction industry and all industries by age group, 2016 to 2019

Age group	Construction industry – no. of fatalities	Construction industry – average death rate per 100,000 FTEs	All industries – average death rate per 100,000 FTEs
Under 20	13	7.8	1.48
20–24	63	8.1	2.15
25–34	187	7.5	2.55
35–44	214	7.7	2.96
45–54	241	9.6	3.4
55–64	229	12.7	4.41
65 & over	**100**	**22**	**7.89**

Source: Centers for Disease Control and Prevention (11)

Table 3.7 Worker fatalities in the US: Construction industry accident by mechanism, 2016 to 2019 (combined total)

Construction accident mechanism	No. of fatalities	% of fatalities
Fall to lower level	1461	54%
Struck-by	691	25.5%
Electrocution	319	11.8%
Caught-in/between	237	8.7%

Source: Brown, Harris (10)

3.4   WHS Performance of the UK Construction Industry Compared to That of Other Industries

According to the HSE (12), the construction industry includes three broad categories:

- Construction of buildings – general construction of buildings, including new work, repair, additions and alterations;
- Civil engineering – civil engineering work, including road and railway construction and utility projects; and
- Specialised construction activities – covering trades that usually specialise in one aspect, common to different structures; for example, demolition, electrical, plumbing, joinery, plastering, painting and glazing.

In the UK, a new era started with mandatory Construction Design and Management (CDM) instruction in 1994 and the integration of construction workers' safety into design. According to this instruction, designers are obliged to avoid all foreseeable hazards at the construction stage. This regulation clearly recognised the designers' responsibility towards construction workers' health and safety. CDM instructions oblige designers to eliminate the hazard at its source. If that elimination is not possible, substitution of a safer practice is another option. However, if the risk remains, information regarding risk characteristics could help the construction team to devise more accurate solutions via engineering or administrative controls (13). One of the prime aims of CDM instruction is to promote a collaborative approach in the industry, especially among the design group, by taking into account all who are participating in the design and construction process to consider health and safety issues. The CDM 2015 instruction makes it mandatory to employ a CDM coordinator from the very early stage of the design to have control over safety practices from the beginning of each project (14).

The annual population survey of Great Britain indicates that in the financial year 2020/21, the construction sector employed 2.08 million individuals, accounting for

6% of the workforce in the country (12). In the same period, 39 workers and four members of the public were fatally injured. Between 2014/15 and 2020/21, 180 work-related deaths were reported in the UK's construction industry, representing an average of 39 deaths per year. Furthermore, the construction industry ranks third among all industries in terms of occupational fatalities only after agriculture, forestry and fishing and water supply, sewerage, waste management and remediation activities. Most of these incidents (95%) occurred in the construction of buildings within the specialised construction activities sub-division (Table 3.8). The construction industry ranks sixth among all industries in terms of non-fatal occupational injuries. As for fatal injuries, the construction of buildings within the specialised construction activities sub-division had the highest non-fatal injury rate of 309 per 100,000 workers (Table 3.9).

In terms of non-fatal injuries, between 2018/19 and 2020/21, falling from height was the main mechanism in construction work. Also, (15) reports the accident types of 'specified injuries' which are:

- fractures, other than to fingers, thumbs and toes;
- amputations;
- any injury likely to lead to permanent loss of sight or reduction in sight;
- any crush injury to the head or torso causing damage to the brain or internal organs;
- serious burns (including scalding) which:

  cover more than 10% of the body or cause significant damage to the eyes, respiratory system or other vital organs;

- any scalping requiring hospital treatment;
- any loss of consciousness caused by head injury or asphyxia;
- any other injury arising from working in an enclosed space which:

  leads to hypothermia or heat-induced illness or requires resuscitation or admittance to hospital for more than 24 hours (Table 3.10).

Table 3.8 Average worker fatalities in the UK: Construction industry occupation groups, 2014/15 to 2020/21

Occupation	No. of fatalities	No. of fatal injury per 100,000 workers
Construction	39	1.91
Construction of buildings; specialised construction activities	37	2.18
Civil engineering	**2**	**0.58**

Source: RIDDOR (15)

Table 3.9 Average worker non-fatal injuries in the UK: Construction industry occupation groups, 2014/15 to 2020/21

Occupation	No. of reported non-fatal injuries	No. of non-fatal injuries per 100,000 workers
Construction	3,464	272
Construction of buildings; specialised construction activities	3,041	309
Civil engineering	**423**	**145**

Source: RIDDOR (15)

Table 3.10 Worker non-fatal injuries in the UK: Construction industry accident by mechanism, 2018/19 and 2020/21

Construction accident mechanism	No. of incidents resulting in specified injuries	No. of injuries in total resulting in being away from work for more than 7 days
Injured while handling, lifting or carrying	6%	26%
Slip, trip or fall on same level	31%	23%
Fall from height	33%	11%
Struck by moving (including flying/ falling) objects	14%	11%

Source: Health and Safety Executive (12)

3.5   Cost of Construction Accidents in Australia, the US and the UK

3.5.1   *Assessment of Construction Injury Costs*

A great deal of money is lost each year as a result of construction accidents for the construction organisation, the sector and the nation. Due to the magnitude of the problem determining a definitive cost is often complicated. It is common for economists to advocate for assessing the benefits of investments in health and safety on the basis of a cost-benefit analysis (16). There is a distinctive iceberg effect that characterises the costs associated with accidents. In contrast to the visible and tangible costs associated with the tip of the iceberg, the bulk of the iceberg lies submerged and conceals much of its hidden and often indeterminate costs.

According to (17):

> If the true costs of injuries were well defined, management would be in a better position to make informed decisions concerning safety. Rather than addressing safety solely from an altruistic point of view, owners should also consider safety from a more purely economic perspective.

Many researchers have echoed these sentiments, making it essential to understand the true costs of construction accidents to inform management properly. In recent years, several researchers have attempted to investigate the true cost of construction accidents by identifying the direct and indirect costs associated with such incidents (18–21). Table 3.11 presents the various frameworks developed for calculating the true cost of construction injuries.

To assess the direct and indirect costs of construction injuries, several studies have employed a ratio-based approach. Recently, workplace health and safety authorities in several countries have adopted robust methods to determine the cost of construction injuries by adding the direct and indirect costs into their calculations. Based on an estimate from Heinrich (22), indirect costs are approximately four times the direct costs in construction accidents. A ratio of 10:1 was reported by Sheriff (23), and a ratio of 50:1 was reported by Bird and Loftus (24). Despite the wide variation in research conclusions, construction managers commonly use ratio-based approaches to estimate the total cost of accidents to a preliminary degree (25). Nevertheless, Gavious, Mizrahi (26) argue that managers tend to underestimate real accident costs using the ratio-based approach because they assume that the majority of expenses are covered by insurance and do not devote much time to calculation methods that require extensive data collection.

Workers' compensation includes payments to employees, including wages while they're not fit for work, medical expenses and rehabilitation, which can be considered direct costs of the injuries. The following sections analyse the latest available work-related injury compensation statistics of three developed countries, Australia, the UK and the US, along with available data regarding the true cost of construction injuries to better understand the magnitude of loss.

3.5.2   Cost of Injuries in Australia

According to Safe Work Australia (27), in the financial year of 2017–18, 18,020 serious claims were made by construction workers, which can be translated to 15.3 serious claims per 1,000 employees. Furthermore, the median time lost caused by these injuries is reported to be as high as seven working weeks and the median compensation paid per claim was A\$15,600 for the same period.

According to the most recent study conducted by Safe Work Australia (28), in Australia, the true cost of work-related injuries and illnesses totalled A\$61.8 billion in 2012–13. This figure encompasses injury and illness costs for Australian employers, workers and the community. The results of earlier SWA studies found similar results, with the cost of work-related injuries estimated at \$57.5 billion and \$60.6 billion, respectively, in 2005–06 and 2008–09.

3.5.3   Cost of Injuries in the US

A recent study on the true cost of construction injuries was conducted by Jill Manzo (29) for the accidents that happened from 2011 through 2015. The analysis in this study was based on the cost categories identified by Waehrer, Leigh (30) and

Table 3.11 Frameworks of calculating the cost of construction-related injuries

Author	Categories of injury costs
Everett, Frank Jr (19)	• Direct costs of injuries and fatalities, including workers' compensation and public health costs • Indirect costs associated with injuries and fatalities include loss of productivity, disruption of schedules, administrative time spent conducting investigations and reports, training replacement personnel, wages paid to the injured workers and others for time not worked, clean-up and repair, adverse publicity, third-party liability claims and equipment damage • Costs associated with health and safety programs, including salaries for H&S, medical and clinical personnel, H&S meetings, inspections of tools and equipment, orientation sessions, site inspections, personal protective equipment, health programs and miscellaneous supplies and equipment
Waehrer, Dong (18)	• Direct costs of hospital, physician, nursing home, home health care, medical equipment and burial expenses; insurance-related administrative costs associated with medical claims; payments for mental health treatment; police, fire, emergency transportation and coroner services; and property damage • Indirect costs associated with wages lost to the victim, household production losses and administrative costs incurred by the compensation program such as workers' compensation wage replacement programs and sick leave benefits • Costs related to quality of life that reflect the pain and suffering that victims and their families experience as a result of injury, death or illness
Rikhardsson, Impgaard (20)	• Hours which employees and managers spent on activities related to the accident, as well as those for which wages were paid without work effort in return, including standstill periods in production and employee sick leaves • Costs of materials and components affected by or lost as a result of an accident, such as spare parts for machines, replacement materials for damaged materials and value of products not produced • Temporary replacements, consultants and legal support were purchased as a result of the accident • Financing and rehabilitation costs (incurred less frequently)
Rikhardsson, Impgaard (21)	• Variable costs associated with sick leave, such as sick pay and supplements to full salaries • Occupational accident–related fixed costs that do not depend on the length of absence of affected workers, such as communication and administrative costs • Typical disturbance costs, including production loss, lost time, fines and overtime costs depending on the specific accident and the roles, tasks and competencies of the injured worker/s.

adjusted to constant 2017 dollars. The adjusted average cost of fatal occupational injuries in the private sector was estimated to be $5.34 million. Nationally, the 867.8 average annual construction worker fatalities during this period cost $4.63 billion per year. Accordingly, construction-related deaths cost the United States nearly $5 billion each year in the form of diminished productivity, lost family income, pain and suffering and reduced quality of life.

Regarding non-fatal injuries, Waehrer, Dong (18) analysed the 2002 national incidence data from the BLS to assess the true cost of construction accidents in the US. This study reported that the total direct and indirect costs of non-fatal accidents were approximately $7 billion, while an average non-fatal injury involving days away from work cost roughly $42,000. As for the fatal accidents, this analysis considered lost production, lost family income, pain and suffering costs and reduced quality of life for each year.

3.5.4   Cost of Injuries in the UK

In its latest report, the HSE assessed the true cost of injuries in Great Britain for the financial year of 2018–19. This assessment covered the cost to individuals and their families, as well as employers, the government and broader society. In their assessment approach, financial costs such as lost production and health care expenses are measured in addition to human costs (the impact on an individual's quality of life and, in the case of fatal injuries, the loss of life). The following details are considered in this method:

- "Financial" expenses incurred – either as a result of payments for services or as a result of income or production lost due to injury or illness. A number of factors contribute to these expensive, including productivity, health care, mental health and rehabilitation, administrative and legal and Employers Liability Compulsory Insurance costs.
- "Human costs" – the monetary value of the impact on quality of life and/or the loss of life of affected workers. Often, this is the most significant impact of illness and injury. By estimating these costs in monetary terms, the total economic burden of workplace injuries and illness can be better understood.

(31) evaluated construction statistics from HSE reports for the period from 2018 to 2021 and found that the cost of injuries and illness in the UK construction industry was estimated at approximately £16.2 billion per year. The government incurred 22% (£3.5bn) of those costs, while employers incurred around 20% (£3.16bn). Nevertheless, the majority of these costs (59% or £9.56bn) were borne by injured and ill individuals themselves.

3.6   Summary

An overview of construction health and safety performance indicators was provided in this chapter. Following this, lagging construction health and safety performance indicators were presented and analysed in three developed countries: Australia, the United Kingdom and the United States. Based on these findings, it can be concluded that the WHS performance of the construction industry in all three cases was drastically lower than that of most other industries. Finally, the true costs of occupational accidents were presented using the most recent data for each country. In all cases, the accident costs were extremely high, requiring the governments to take immediate action to implement interventions.

References

1. Chan A, Hon C. *Safety of Repair, Maintenance, Minor Alteration, and Addition (RMAA) Works: A New Focus of Construction Safety*. New York: Routledge; 2016.
2. Australian Constructors Association. *Lead Indicators Safety Measurement in the Construction Industry*. Australia; 2015.
3. Biggs H, Dinsdag D, Kirk P, Cipolla DJ. Safety culture research, lead indicators, and the development of safety effectiveness indicators in the construction sector. *IJoT, Knowledge, Society*. 2010;6(3):133–40.
4. Hinze J, Thurman S, Wehle A. Leading indicators of construction safety performance. *Safety Science*. 2013;51(1):23–8.
5. Safe Work Australia. *Work-Related Injuries and Fatalities in Construction, Australia, 2003 to 2013*. Australia; 2015.
6. Safe Work Australia. *Australian Work Health and Safety Strategy 2012–2022*. Canberra, ACT: Commonwealth of Australia; 2011.
7. Safe Work Australia. *Work-related Traumatic Injury Fatalities*. Australia; 2020.
8. Bureau of Labor Statistics (BLS). *About the Construction Sector*; 2022. Available from: www.bls.gov/iag/tgs/iag23.htm.
9. Bureau of Labor Statistics (BLS). *National Census of Fatal Occupational Injuries in 2016*; 2020. Available from: https://www.bls.gov/iif/home.htm#2018.
10. Brown S, Harris W, Brooks RD, Dong XS. *Fatal Injury Trends in the Construction Industry*. Georgia: Centers for Disease Control and Prevention; 2021.
11. Centers for Disease Control and Prevention. *Work-Related Injury Statistics and Resource Data Systems (WISARDS)*. Georgia: Centers for Disease Control and Prevention; 2022. Available from: https://wwwn.cdc.gov/Wisards/.
12. Health and Safety Executive. *Construction Statistics in Great Britain, 2019*; 2021. Available from: https://www.hse.gov.uk/statistics/industry/construction.pdf.
13. Wright M, Berman G. *The Case for CDM: Better Safer Design: A Pilot Study*. Suffolk: HSE Books; 2003.
14. Health and Safety Executive. *Construction (Design and Management) Regulation UK: Health and Safety Executive*; 2016. Available from: www.hse.gov.uk/.
15. RIDDOR. *Reporting of Injuries, Diseases and Dangerous Occurrences Regulations*; 2020. Available from: https://www.hse.gov.uk/statistics/tables/index.htm.
16. Hjalte K, Norinder A, Persson U, Maraste P. Health – health analysis – an alternative method for economic appraisal of health policy and safety regulation: some empirical Swedish estimates. *Accident Analysis & Prevention*. 2003;35(1):37–46.
17. Hinze J. *Construction Safety* (2nd ed.). Upper Saddle River, New Jersey: Prentice-Hall Inc; 2006.
18. Waehrer GM, Dong XS, Miller T, Haile E, Men Y. Costs of occupational injuries in construction in the United States. *Accident Analysis & Prevention*. 2007;39(6):1258–66.
19. Everett JG, Frank Jr PB. Costs of accidents and injuries to the construction industry. *Journal of Construction Engineering and Management*. 1996;122(2):158–64.

20. Rikhardsson P, Impgaard M, Mogensen B, Melchiorsen A. Virksomhedens Ulykke-somkostninger (The Corporate Costs of Occupational Accidents). *The Aarhus School of Business—PricewaterhouseCoopers.* 2002. Available from: https://bss.au.dk/en/cooperation/partnerportalen/pwc.

21. Rikhardsson PM, Impgaard M. Corporate cost of occupational accidents: an activity-based analysis. *Accident Analysis & Prevention.* 2004;36(2):173–82.

22. Heinrich HW. *Industrial Accident Prevention. A Scientific Approach* (2nd ed.). Monterey, California: McGraw-Hill; 1941.

23. Sheriff R. Loss control comes of age. *Professional Safety.* 1980;25(9):15–18.

24. Bird FE, Loftus RG. *Loss Control Management.* Loganville, GA: Institute Press; 1976.

25. Manuele FA. Accident costs rethinking ratios of indirect to direct costs. *Professional Safety.* 2011;56(1):39.

26. Gavious A, Mizrahi S, Shani Y, Minchuk Y. The costs of industrial accidents for the organization: developing methods and tools for evaluation and cost – benefit analysis of investment in safety. *Journal of Loss Prevention in the Process Industries.* 2009;22(4):434–8.

27. Safe Work Australia. *Australian Workers' Compensation Statistics 2018–19*; 2020. Available from: https://www.safeworkaustralia.gov.au/doc/australian-workers-compensation-statistics-2018-19.

28. Safe Work Australia. *The Cost of Work-related Injury and Illness for Australian Employers, Workers and the Community: 2012–13.* Australia; 2015.

29. Manzo J. *The $5 Billion Cost of Construction Fatalities in the United States: A 50 State Comparison.* Midwest Economic Policy Institute; 2017.

30. Waehrer G, Leigh JP, Cassady D, Miller T. Costs of occupational injury and illness across states. *Journal of Occupational Environmental Medicine.* 2004;46:1084–95.

31. Herts Tools. *Accidents in the Construction Industry: Report*; 2021. Available from: https://hertstools.co.uk/accidents-in-the-construction-industry-report/.

4 Accident Causation Models and WHS Challenges in Construction Projects

4.1 Occupational Accidents

4.1.1 Introduction

Several accident models have been developed to facilitate the identification of shortcomings and failures leading to accidents. A review of accident models by Katsakiori, Sakellaropoulos (1) indicated a gradual paradigm shift over time from looking for a singular cause of accidents to recognising a chain of causes and categorised such models into three groups, as follows:

- Sequential accident models, presenting an accident as a series of failures in a specific circumstance
- Human information processing models, presenting an accident as the result of unsafe behaviour and activities
- Systemic accident models, presenting factors related to the performance of organisation and management systems as a whole

4.1.2 Sequential Models

Models such as the "domino model", which describes clear-cut and direct cause-and-effect relationships among multiple failures, are considered sequential models (1). In the domino model, an accident is perceived as the result of a chain of actions initiated by an individual's personal condition and surrounding environment that lead to a loss-producing event (2). Bird and Loftus (3) updated this model by introducing the role of managerial controls as causal factors of accidents. As a result, whether an accident causes bodily injury, sickness, or loss of property, managerial controls are to blame. The accident, evolution and barrier model developed more recently examines accidents as consequences of failures in a chain of individual barriers (4). Hollnagel (5) and Goh, Brown (6) argued that sequential models are restricted to immediate accident circumstances and fail to describe accident causations in complex situations, such as industrial accidents.

DOI: 10.1201/9781003224853-4

4.1.3 Human Information Processing Models

Human information processing models present an accident as a result of unsafe behaviours and activities. Gordon, Flin (7) argued that a solid approach towards documentation and analysis of physiological and human factors involved in accidents was lacking and developed the Human Factors Investigation Tool (HFIT). This tool categorises accident causations into four groups: 1) unsafe behaviours before the accident, 2) a psychological state of mind produced the unsafe behaviours, 3) a work environment condition that leads to such a psychological state of mind and 4) behavioural responses that could prevent such an error. As part of HFIT, unsafe behaviours have been included to allow for a more in-depth understanding of the nature of the error before investigating the cause. This model defines the psychological state as the detection cue with four elements: Internal feedback, system feedback, external communication and planning behaviours. The third element of the HFIT involves situation awareness, defined by Endsley and Garland (8) as the ability to perceive, comprehend and project the status of the elements of an environment within a space and time. The last component of this model refers to responses to situations that may encourage the occurrence of errors.

4.1.4 Systemic Accident Causation Models

Unlike sequential models, systemic accident causation models describe the existence of dynamic interaction among cultural and organisational factors in creating a hazardous situation (9). The "Swiss cheese model", developed by James Reason, is one of the popular models in this category (2). According to the Swiss cheese model, accidents are triggered by a complex interaction between active failures and latent conditions. As Heinrich's unsafe acts or conditions model mentions, active failures are apparent causes and can be recognised easily (10). However, the latent condition requires expert investigation and is considered an "accident waiting to happen" (9). For example, design errors, poor training and compatibility problems between workers and their given responsibility fall into this category. A combination of latent conditions and human presence in industrial works leads to a human violation or unsafe acts (1). However, Reason (11) indicated that among many unsafe acts, only a few lead to accidents due to the in-built defence of the system. Similarly, in the Swiss cheese model, layers act as barriers and have holes that vary in size over time. The system defence stops in unfortunate situations where holes are positioned in the same direction (2). Rasmussen and Vicente (12) pointed out that existing constraints related to objectives in the work system, such as a compact work schedule, financial considerations and safety and functional obligations, influence acceptable safe behaviour. In the model described by Rasmussen and Vicente (12), a mixture of natural human willingness to find means to spend less effort to deliver a job and demands for efficient production lead workers to push the boundaries of acceptable behaviour. This means that if a working system continuously presses workers to work close to the boundaries of acceptable behaviour, such a system should expect consequent errors.

4.2   Construction Accident Models

In the past, a large number of studies on construction accidents were developed based on compensation-based accident reports (13–17). Such reports utilised data that were poor in quality and/or incomplete. Daniels and Marlow (18) analysed the level of completeness of non-fatal construction accident reports and found that only 46% of the relevant information was provided. Studies in the UK found construction companies' accident reports to be poor in quality, making it difficult to undertake systematic accident causation analysis (19, 20). The primary focus of early construction accident studies was on the immediate circumstances, such as unsafe behaviour and attributes, as the leading causes of accidents. For instance, Hinze (21) recognised distraction as a predominant cause of construction accidents, while Abdelhamid and Everett (22) discussed training, safety behaviour and work schedule, and Whittington, Livingston (23) rationalised the influence of management, workplace conditions and individual worker characteristics on the cause of accidents as 1:2:1. Suraji, Duff (24) analysed distal and proximal accident causes in construction projects and distinguished worker behaviours, site settings and construction methods and practices as proximal causes, with managerial influence on the project process as a distal cause of accidents.

The importance of distal factors has found greater acknowledgment in recent studies. Priemus and Ale (25) analysed the Bos and Lommerplein estate project in Amsterdam using James Reason's model (11) and found a serious failure of barriers at different stages of design, inspection and construction. In a similar study, Manu, Ankrah (26) analysed how different characteristics of construction projects, such as project type, construction method, site condition, project schedule, design errors, sub-contractor-related issues and procurement processes, contributed to accidents and classified them into two groups: Proximal and distal factors. In addition, Manu, Ankrah (26) concluded that a large number of these factors were rooted in client requirements, design decisions and project management decisions.

Mitropoulos, Abdelhamid (27) proposed an accident causation model based on Rasmussen's work (12) considering the hazards created as a result of combining construction activities and context characteristics. Mitropoulos, Abdelhamid (27) suggested that exposure to hazardous situations can be managed by the workers' decisions regarding how to control and handle a give situation. In the case where the worker fails to control the hazard, there is potential for an accident.

The Human Factors Analysis and Classification Scheme (HFACS), primarily developed for the aviation industry, was adapted by Hale, Walker (28) to analyse construction accidents. The customised HFACS for construction is structured with two primary classes of unsafe acts: Deliberate (violations) and errors, and each class is further divided in detail. Deliberate unsafe acts are sub-divided into routine, situational and exceptional, and errors can be related to knowledge, rules or skill. The HFACS framework identifies a sequence of cause-and-effect relations between unsafe acts, preconditions and environmental or organisational influences. Data collection for the framework was based on structured interviews with government health and safety inspectors to identify fatal accident causations. The developed framework was validated through real case studies and demonstrated a significant contribution of extending accident causations at the organisational level (28). Garrett and Teizer (29) also employed the HFACS to analyse occupational errors and educate construction workers.

Gibb, Haslam (30) analysed 100 primarily minor construction accidents within a three-year research program at Loughborough University and developed a hierarchical ConAC framework to facilitate in identifying the chains of events leading to unfortunate incidents. The ConAC framework became a popular method due to the real-time nature of the primary data and the ability to collect possible details from interviews and focus groups with the victim(s) and witnesses, along with site observations and photographs. The ConAC model is presented in three levels: Immediate accident circumstances, shaping factors and originating influences, as depicted in Figure 4.1 (9). Figure 4.1 also shows the casual relationships

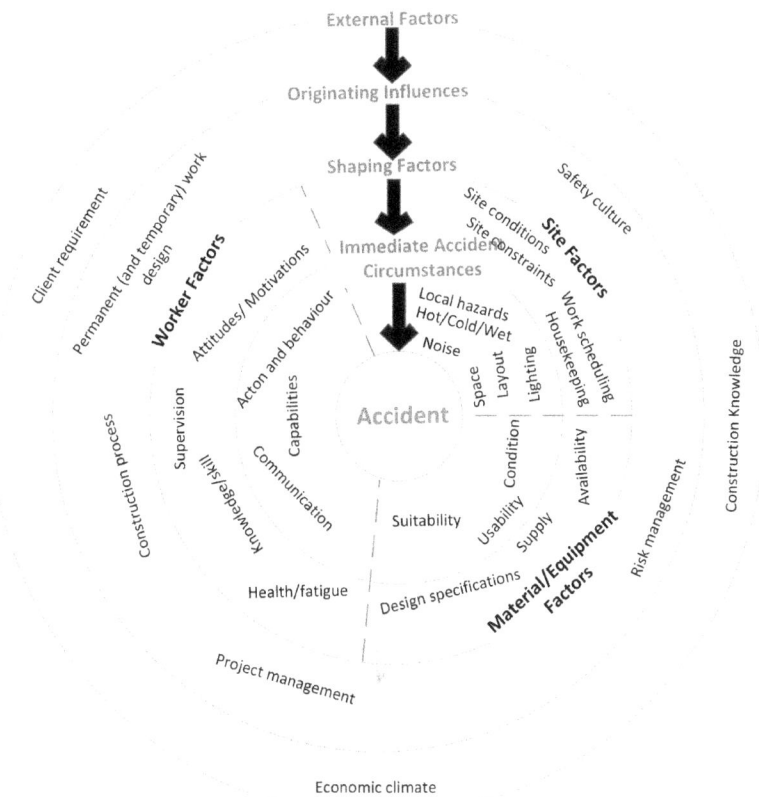

Figure 4.1 ConAC accident causation model

between the factors of each level, starting from originating influences and ending with three immediate accident circumstance groups: The work team, workplace and materials/equipment. The model presents a network of causations that can lead to the occurrence of accidents. Based upon Reason's (11) Swiss cheese model, the ConAC model depicts a more complex interrelation among the causations, allowing for network analysis among the actors (9).

4.3   Construction Accident Risk Drivers

4.3.1   Immediate Accident Circumstances

One explanation for the high rate of fatalities in construction accidents could be found in the highly human-dependant nature of the construction industry compared to, for example, the automotive industry. Despite all of the efforts made through training sessions, construction workers remain prone to making errors. Phua (31) analysed a sample of 2,000 UK citizens employed in a range of different occupations and concluded that physical risk-taking propensity among construction workers was significant.

Alsamadani, Hallowell (32) noted the importance of safety-related communications among crew members for safe performance. This factor is even more significant among workers with different speaking languages. In the US construction industry, the rate of fatalities in the period of 2008–2010 among Spanish-speaking construction workers was 6% higher than non-Spanish-speaking workers (33). In Germany, Arndt, Rothenbacher (34) found the likelihood of fatal injuries caused by falling objects to be four times higher for foreigner workers than native workers. Alsamadani, Hallowell (32) explained that the language barrier was a potential reason for the difficulty in exchanging safety knowledge among workers and for such disproportionate injury rates.

Construction projects also involve workers from different cultures working together. Comu, Unsal (35) assessed the impact of cultural diversity on a project network's performance and found a counter-productive effect at the initial stages; however, such diversity was beneficial to long-term performance. It is challenging to effectively execute a safety program when there is cultural diversity among crew members (32). Vecchio-Sadus (36) explained the necessary considerations for managers to improve the safety communication of workers as:

• ensuring transparent and open conversation about safety concerns with all workers, even with their own language;
• using a primary language within the crew to communicate;
• giving spoken notices to improve safe behaviour; and
• providing workers with a simple and straightforward description of workplace safety.

Field-level hazard recognition performance largely depends on the ability of workers to detect relevant hazards in the middle of irrelevant distractions. However, recent research suggests that workers lack essential safety skills despite having received substantial training (37). Not surprisingly, a large number of injury investigation reports reveal deficits in the hazard recognition ability of workers and recommend more training to prevent injury reoccurrence (10, 38). When hazards remain unrecognised, the likelihood of injuries increases substantially (39). Moreover, the assumption that workers are skilful enough to predict future work conditions and anticipate hazards was shown to be inaccurate in practice (40). For instance, Mitropoulos, Abdelhamid (27) pointed out that workers have difficulty visualising and predicting future tasks due to the dynamic nature of construction projects.

In order to examine the hazard recognition skill of workers, Jeelani, Albert (41) exposed construction workers to photos from several accident cases. They found that workers' selective attention or inattention, the presence of an unknown set of potential hazards and the perception of workers that specific hazards imposed low or minimal levels of safety risks were the most common cause of unrecognised hazards. They realised that several hazards not associated with the primary task being performed remained unrecognised in the case images. These hazards existed within the work environment, such as failure to recognise the presence of overhead power lines in one case image, but were either secondary or unrelated to the primary task being performed (Jeelani et al. 2017).

Studies by Hinze (42) and Haslam, Hide (38) found that around 70% of construction accidents involved the unsafe behaviour of workers. Fam, Nikoomaram (43) defined unsafe behaviour as any action of a worker neglecting safety instructions, procedures, standards, rules and specified criteria of an organisation that has the potential to affect the safety system negatively or put a worker or their colleague in danger. Hallowell, van Boven (44) and Namian, Albert (45) found such unsafe behaviours to be a result of poor training and worker hazard identification skills. (46) found that the traditional safety training approach failed to provide essential safety knowledge for workers. Mohammadfam, Ghasemi (47) analysed workers' behaviour using a Bayesian network and found safety attitude, safety knowledge and a supportive environment to be the best intervention factors to improve safety behaviour.

The dynamic nature of construction sites causes time-to-time changes that do not occur in fixed industrial facilities; site layout and structures vary during different stages of the projects, and sites are affected by the surrounding environment as well as by climatic changes. Toole (48) described the importance of proper site conditions as a factor that could decrease the chance of an accident occurring. He pointed out that some of these conditions included poor housekeeping, slippery surfaces, excessive accumulation of trade workers in one location, overhead and underground electrical lines and weather conditions. As a concept, when considering accidents, a circumstance of loss of control to which victims are exposed, it can be assumed that without exposure, no accident would take place (48).

Yi, Chan (49) indicated that extreme temperature, noise, vibration, dust, radiation and chemicals were examples of health hazards that construction workers could be exposed to. Although no comprehensive study has accurately measured the health impacts of construction activities, they can be classified into short-term (acute) and long-term (chronic) illnesses depending on the exposed environment. Heat stress is a well-known environmental hazard which causes both psychological and physical discomfort in workers depending on the frequency and intensity of the exposure, along with declined productivity, amplified chanced of an accident and even threats to the life of workers (50–52). Taylor (53) explained that perceptual, cardiovascular and thermoregulatory strains increase the possibility of emotional stresses, such as irritability and confusion, which in turn cause workers to underestimate potential hazards. Horie (54) reported that in 2010, 47 occupational fatalities occurred in Japan in different industries due to heat stroke. Petitti, Harlan (55) highlighted a higher chance of construction workers being exposed to extreme weather-related injuries and fatalities. Yi, Chan (49) presented the results of a survey in Hong Kong related to construction workers exposed to hot weather, indicating that 5% of all construction workers suffered heat stroke and 23% were subjected to heat stroke symptoms. Joubert, Thomsen (56) also explained how severe construction environments could be in regions such as the Middle East that regularly experience 90% humidity and 45 °C temperatures.

Huang and Wong (57) argued that a safer working environment should be provided for construction workers through safe design of the site layout. Toole (48) suggested three steps to prevent accidents rooted in poor site conditions:

1) An entity must understand the appropriate and inappropriate conditions for performing each task.
2) Considering the dynamic environment of the construction site, an entity must monitor the actual setting of the site and ongoing activities considering hidden hazards and perform a risk analysis based on drawings or other references.
3) An entity must consider either controlling the performed activity or site condition according to the risk analysis.

Construction materials are generally chosen from those that are safe to use. However, due to some structural requirements (durability, strength etc.) or client requirements (cost, time, design etc.), it may be necessary to use harmful materials with a great deal of safety precautions. A small spill of harmful materials during transport, storage or handling can bring about acute or chronic disorders in workers. Payne, Van Valkenburgh (58) discussed the importance of safe storage of materials such as paint-related products, lubricants and solvents and proposed a mobile structure for safely handling such materials. Using asbestos as a construction material is banned in many countries due to resulting health issues (59, 60). However, the wide use of asbestos-based products in the past remains a concern, as maintenance trade workers can be exposed to health hazards of this type. Hot mix asphalt is a widely used construction material that emits fumes during its placement and compaction. Hazardous chemicals inside asphalt fumes, such as carbon monoxide, nitrogen oxides, sulphur, particulates, polycyclic aromatic hydrocarbons and volatile organic compounds can cause serious health problems for workers (61).

Moving from a worker-based to a semi-automated execution of construction projects requires the involvement of plants, machinery and equipment. D. Holt and J. Edwards (62) explained the existence of substantial hazards, such as struck-by injuries for operators and workers near machinery and plants. Edwards and Love (63) outlined the hazards involved in using mini-excavators, such as unintentional activation, struck-by injuries during lifting or load detachment, insufficient visibility leading to crushing and/or struck-by injuries, hitch displacement, instability and service strikes. Holt (64) explained the reason for such hazard as being that almost all construction plants and machinery are designed and manufactured by manufacturers who are not construction experts.

4.3.2   Shaping Factors

Several studies have explained the differences in the attitude of workers regarding taking risks in their operation (65–67). Brown, Willis (65) predicted that subunit and/or perceived organisational stress to prioritise expediency over safety behaviour can increase the chance of justified arrogance and/or a willingness to rationalise risk-taking actions among workers. DeJoy (68) indicated that when barriers related to a job cause difficulty in safe execution, this influences the worker's attitude towards following safety guidelines and rules. This situation worsens when the worker realises that managers do not pay enough attention to preventing hazards and conveys the message to the worker that their safety is not a priority for the organisation (68).

Site supervisors facilitate and control safety behaviour among the work team using national and local WHS regulations. A recent study compared accident causations between the neighbouring countries of Sweden and Denmark, which have similar regulations, and showed the significant impact that site supervisory practices can have on reducing the number of occurred accidents (Nielsen, 2016). There is no clear definition regarding what can be considered the most efficient and effective method for site inspection within the existing literature (69, 70).

The physical and mental readiness of workers is key to executing the work in a proper manner. Grandjean (71) described workers' fatigue as a combination of various factors, such as environmental factors (noise, light, air at the site, etc.), personal factors (lifestyle, sex, age, etc.) and occupational factors (working hours, stress, the task, etc.). Such a combination results in physical fatigue, such as pain related to excessive muscle stress and mental fatigue, such as sensation weariness (71).

Uher and Zantis (72) defined planning and scheduling as processes of predicting future activities and outcomes that could be indefinite or unknown. Thus, devising a suitable course of action requires assessment of future activities and access to sufficient provisions based upon opinions and facts. Although past research on construction scheduling has primarily concentrated on optimising resource allocation (e.g., Bakry, Moselhi (73) Lucko (74)) and the cost of projects (e.g., Hegazy and Kamarah (75)), safety considerations have drawn attention in recent years (76–78). Consideration of the WHS regulations in the project planning phase and safety risk analysis of defined activities have been the two main research topics in construction safety. For example, Kartam (79) and Hinze, Nelson (80) designed a tool/framework to embed safety knowledge in the critical path method schedules. Yi and Langford (81) and Wang, Liu (82) integrated safety risk with a project schedule to determine the work period with the highest possibility of accidents and alleviate such hazards using suitable planning.

Bluff (83) argued that minimising or alleviating a large number of safety hazards related to construction equipment operations could be possible through the supply chain and manufacturers' initiatives. Such initiatives could consider safety measures in the design specifications of construction materials and equipment and replace them in the market instead of producing unsafe materials. Bluff (83) also classified such safety measures within the literature into two types – "safe place" and "safe person" measures. Safe place controls reduce or eliminate hazards by embedding physical safeguards between people and any produced or transferred forms of energy (84). Safe person controls involve those systems continually alerting operators, maintenance workers, or people who may encounter a safety risk. Well-known examples in this category are safety signs and personal protective equipment (PPE).

4.3.3 Originating Influences

Owners, contractors and designers all have the opportunity to make early decisions in terms of improving construction safety. According to (85), the likelihood of controlling and eliminating construction hazards is rapidly lost as the project matures and moves from the design to the construction phase. It is, therefore, ideal to understand the constructive implications of the inception, concept design and detailed design.

Behm (86) studied 224 fatal accident circumstances explained in the database of the National Institute of Occupational Safety and Health and concluded that 42% of the accident causations could have been designed out before construction began. Another study by Gibb, Haslam (87) reviewed 100 construction accidents and found that in 47% of the cases, changes in the permanent design could have reduced the likelihood of accidents. One study by the National Occupational Health and Safety Commission (NOHSC) of Australia found that between 2000 and 2002, nearly half (44%) of work-related fatalities in the construction industry had roots in design (88). Studies have also confirmed a similar correlation between design errors and rates of construction fatalities (63% correlation in Europe (89) and 36% in the UK). Meanwhile, a study by Tymvios, Gambatese (90) revealed a significant lack of WHS knowledge among the design team in 5.4% of architects and 19.3% of engineers when considering safety measures in their designs.

The earliest construction worker safety studies were initiated by Dr Jimmie Hinze and Francis Wiegand in 1992 and concentrated on the influence of designers on construction worker safety. They contended that, "despite the obvious reasons for placing the primary responsibility on the contractor, the safety performance of a project can be excelled to a larger extent through decisions made by the designer" (91). Following that study, a well-formed idea of designing for construction worker safety was initiated, despite the existing notion that designers simply had no responsibility for the safe completion of a project (92).

In the case of the UK, a new era started due to mandatory construction design and management (CDM) instruction in 1994 that integrated construction workers' safety into design. According to this instruction, designers are obliged to avoid all foreseeable hazards at the construction stage. This regulation clearly recognises the designers' responsibility towards construction workers' health and safety. CDM instruction obliges designers to eliminate the hazard at its source. Where elimination is not possible, the substitution of a safer practice is another option (93). However, if the risk remains, information regarding the characteristics that could help the construction team devise more accurate solutions via engineering or administrative controls should be provided (94). One of the prime aims of CDM instruction is to promote a collaborative approach in the industry, especially in the design group, by taking into account all who are participating in the design and construction process when considering health and safety issues. The CDM 2015 update makes it mandatory to employ a CDM coordinator from the very early stage of design to have control over safety practices from the beginning of the project (93).

In Australia, WorkCover NSW publishes construction hazard assessment implication review (CHAIR) checklists to assist designers, contractors, clients and other key stakeholders in collaborating to reducing construction, maintenance, repair and demolition safety risks associated with design (95). CHAIR consists of three stages of design review: One at the conceptual phase of design, followed by a second review that focuses on construction and demolition issues performed prior to construction and a third review at the same time as the second review that focuses on maintenance and repair issues (95).

Pinion, Brewer (96) described the strength of the existing safety climate within an organisation as a leading indicator. The safety climate within an organisation is the safety perception shared by employees and imposed by management (97). The maintenance of a safer climate is facilitated through site supervisors as a part of the managerial system (98). Pinion, Brewer (96) indicated that management commitment is the most important aspect of developing a safety climate among workers. In a situation where management allocates time for employee safety programs and inspection, safety is perceived as the main concern. Toole (48) conducted a preliminary study of the factors contributing to the safety perception of employees within construction companies and the results showed that management commitment had the greatest impact.

4.4   Investigation of Construction Accidents

4.4.1   Overview

Accident models published by the Australian government and found within published literature were investigated to determine the areas that require improvement. The existing models were found to either be outdated (10, 99, 100) or to not provide sufficient information about the root causes of the accidents (101). In addition, adopting OHS models from other countries could be questionable since the roots of accidents more or less vary among countries due to dissimilarities in labour markets, OHS legislative settings and industrial arrangements (10, 102).

Therefore, as a first step, the causes of construction accidents were investigated using a proper accident model. The literature review presented in this chapter was conducted to find accident models developed in the past and establish a suitable method for the case of the construction industry. The ConAC method developed by a research group supported by the HSE of the UK to be employed for construction accident investigations (30) was found to be a suitable model. In 2003, a research team centred at Loughborough University analysed 100 construction accidents and developed the hierarchical ConAC framework to facilitate the identification of why and how construction accidents happen (30). During a three-year period, the research group collected descriptive information related to construction accidents from victims and witnesses of the accidents (e.g. site workers, supervisors, etc.) to recognise the chain of failures leading to construction accidents (Gibb et al, 2006). Several world-famous safety researchers have exploited this model to determine construction accident causations within the context of their own countries (9, 10, 99, 100, 103).

4.4.2   Data Collection

A large number of construction OHS accident causation investigations have been undertaken based on opinion-based risk data retrieved from questionnaire surveys and rated values by experts (104). However, these types of data are subjective and involve biases that can influence the judgments of experts, such as overconfidence, anchoring, availability, representativeness, unrecognised limits or conservatism (105–107). In addition, studies by Gustafsod (108) on gender and by Tixier, Hallowell (109) on how emotional state influences risk perception have shown evidence of bias in such studies. Although there are means to minimise the negative impact of such biases (110), empirical and evidence-based data collection provides a more realistic basis for understanding a targeted situation. Therefore, this study was based on case reports of recent work-related fatal construction accidents. Case reports from the National Coronial Information System (NCIS) of Australia were identified as valuable sources of information due to the richness and availability of the stored data.

According to the NCIS coding manual and user guide (111), mainstream NCIS data is obtained from coronial files and includes police investigation reports, coronial findings, autopsy reports and forensic medical reports (e.g., toxicology). However, it does not include witness testimonies, photograph support or the original transcripts of the inquests.

NCIS cases are categorised according to the following information:

1) Case demographics, including name, age, date of birth, date of death, place of usual residence, marital status, employment status, usual occupation, country of birth, years in the country and indigenous origin
2) Cause of death details, including medical cause of death, mechanism of injury, object or substance producing injury, relationship of the perpetrator to the

deceased, ICD 10 cause of death codes, incident information, date and time of incident and death, as well as when the person was last known to be alive, where the deceased's body was located and activity at the time of death or incident leading to death. In addition, the deceased's occupation and industry are stated when the person was involved in working for income.

3) Classification, including case type (natural, external and body not recovered) and the presumed intent (intentional self-harm, unintentional, assault, complications of medical or surgical care, etc.)

4) Documentation includes police, autopsy and toxicology reports and the coronial findings.

To initiate the search for the NCIS cases, a time frame from January 2007 to December 2016 was selected for an overall period of 10 years. The access provided by the NCIS gave permission to search for closed cases, and the latest closed case at the time of the search was in 2016; therefore, a 10-year period was set from 2007 to 2016.

In the next step, the following selection criteria were applied in the query design of the NCIS:

1) Case status set to be closed
2) Case jurisdiction (set to include all states and territories of Australia)
3) Work-related and unlikely-to-be-known accident causations
4) Incident industry as construction according to ANZSIC code definitions
5) Results including case details, mechanisms, time of the accident and all the reports

Analysis of data meeting these criteria allowed for more detailed assessment of fatal accident causes than would otherwise be possible through an analysis of national compensation-based statistics.

4.4.3 *Deductive Thematic Content Analysis*

Qualitative, descriptive methods not only originate the knowledge from the collected data but also treat data as living entities to extract factual findings from them (112, 113). According to Vaismoradi, Turunen (114) qualitative descriptive methods, such as thematic analysis and content analysis are more suitable for research that requires less interpretation, in contrast to neuritic phenomenology and grounded theory, which involve a higher level of complexity in interpretations. Due to the nature of the evidence-based data collected in this research (NCIS case reports) and the ability to extract sufficient information with a low level of interpretation, both thematic analysis and content analysis were chosen as research methods for this part of the study. The thematic analysis attempted to determine, analyse and report the holistic patterns (themes) in the data (115). On the other hand, content analysis involving systematic categorisation and coding methods was used to explore the textual information embedded in data to determine trends in words, along with their relationship and structures (114).

There are two modalities for a thematic content analysis within the literature: Inductive and deductive approaches. While an inductive approach deals with data because no previous study deals with the phenomenon, a deductive approach examines a previous theory in a different situation or compares findings for different time periods (116, 117). Deductive thematic content analysis was conducted in this study by coding the collected fatal accident reports collected from the NCIS. The process was facilitated by identifying the ConAC causations. In a technical report for the US Department of Transportation, Behm (118) elaborated on the terminologies of the ConAC model in more detail. Table 4.1 was developed by combining these elaborations with the literature findings mentioned in the previous chapter on accident causation. The model was divided into five major categories: Work team, workplace, equipment, material and originating factors.

Table 4.1 Loughborough University ConAC model terminologies

Group	Code	Term	Explanation
Work team	WT1	Worker actions and behaviours (including attitudes and motivation)	Workers' unsafe actions and behaviours are the last link to complete the chain of events leading to an incident and represent the last chance to control it.
	WT2	Worker capabilities (including knowledge/skills)	This includes both the skills and knowledge of workers for safely conducting a job, using equipment, identifying risks and hazards involved in the job, etc.
	WT3	Communication	Safety communication failures are mainly grounded in poor skills of foreign workers in the country's spoken language, the unwillingness of workers or supervisors to communicate, etc.
	WT4	Immediate supervision	The supervisors, as the traditional controlling members of the management team, are considered well-educated and well-trained on safety issues and have a great deal of influence in identifying and controlling unsafe onsite conditions.
	WT5	Worker health/fatigue	Health issues and fatigue can result in workers' poor concentration, unsafe decision-making and long-term ergonomic problems. This can be the result of prolonged working hours without a break or having no day off for a long time.

(Continued)

Table 4.1 (Continued)

Group	Code	Term	Explanation
	WP1	Site conditions (excluding equipment, materials and weather)	The ground conditions of the site's location and the site's relationship to the surrounding area could contribute to incidents. For example, one condition could be overhead or underground power lines that are already in existence at the site.
	WP2	Site constraints, site layout/space	This includes arrangements at the worksite regarding the given space for safely conducting a task, the relationships of the working team and equipment within the given space.
	WP3	Local hazards	Risks or hazards that are unique to the worksite should be identified and controlled.
	WP4	Working environment (lighting/noise/hot/cold/wet)	Climatic and environmental conditions such as noise, moisture, lack of daylight, etc. could magnify the risk of hazards.
	WP5	Work scheduling	Unsafe work scheduling could result in prolonged working hours, poor preparation and arrangement of the site, an unsafe sequence of tasks, etc.
	WP6	Housekeeping	Risks and hazards are more likely to occur due to an unclean and disorderly placing of machinery, equipment and materials.
Materials	M1	Suitability of materials	Materials should be suitable for a given task. The supplier often provides materials that are considered suitable (by the supplier) for different task elements.
	M2	Usability of materials	The materials utilised should be usable and available.
	M3	Condition of materials	The materials utilised should be in a "safe-to-use" condition. Although new materials are usually safe to use, old and insufficient materials can cause accidents in the maintenance and renovation of buildings.
Equipment	E1	Suitability of equipment	Equipment should be suitable for a task. The supplier often provides equipment that is considered suitable (by the supplier) for different activities.
	E2	Usability of equipment	Usability of equipment includes the equipment being available where it is needed.
	E3	Condition of equipment	The condition of equipment relates to maintaining the equipment so it is in a "safe-to-use" condition.

(Continued)

Table 4.1 (Continued)

Group	Code	Term	Explanation
Originating influences	OI1	Permanent (and temporary) work design	Safe design includes many design elements, such as safer materials, more prefabrication, built-in railings in openings, etc. It also refers to the safe design of temporary structures such as scaffolding and formwork. In addition, information is provided as part of safe design regarding foreseeable hazards, such as underground utilities or overhead power lines, to manage these hazards during the construction phase.
	OI2	Project management	Project management utilises a highly experienced and educated individual or team that can safely oversee and manage unsafe situations by making arrangements regarding contractors and sub-contractors, task sequences, labour supply, etc.
	OI3	Construction processes	Construction processes include method statements, as well as communication of the process among the people involved and appropriate planning based on this process.
	OI4	Safety culture	Safety culture refers to organisation-level attempts to avoid unsafe shortcuts, proper communication of safety concerns among team members, reporting of unsafe situations, implementation of safety measures, etc.
	OI5	Risk management	Risk management includes the process of recognising safety risks, assessing them and taking appropriate control measures.
	OI6	External factors	The influences of external factors are hard to investigate in incidents and are often specific to the regional context. They include the economic climate, construction knowledge and client requirements.

Source: Golizadeh, Hon (103)

4.4.4   *Data Collection from the NCIS Database*

After performing the search, a total of 287 cases were collected. Because all autopsy, police and forensic medical reports were also addressed within the coronial findings, the main source of data for the current study was the coronial finding reports. In the next step, those cases that did not include a coronial investigation were eliminated. This was necessary because police reports only provided details about the location and time of the accidents, while the coronial reports presented detailed investigations about the circumstances of the accident. Searching to remove suicide cases was not required because the granted access to the NCIS database automatically excluded those types of accidents. Accidents that took place in non-construction situations, such as a car accident while commuting or at a mine site, were also excluded from the list.

A total of 105 fatal accident cases with a full coroner's investigation report were collected. Table 4.2 presents the breakdown of fatal cases according to the state/territory and the accident mechanisms. The highest number of fatal accidents were attributed to falling from a height (32% of accidents) and contact with electricity (17% of accidents). Vehicle collisions and being hit by falling objects had the lowest frequencies in the list, appearing in approximately 4% and 8% of the selected cases, respectively. Regarding the states/territories, Queensland (43 accidents) and Victoria (40 accidents) had the highest numbers of accidents. Due to the selection criterion of having a coronial investigation report in the NCIS database, however, most New South Wales cases and all South Australia, Northern Territory and Australian Capital Territory cases were deleted from the list.

More than half of the selected cases took place before 2010 and only 15% of the accidents occurred in the time frame of 2014 to 2016.

4.4.5   *Thematic Content Analysis*

Using the terminologies of the Loughborough ConAC model presented in Table 3.2, the coronial investigation reports of the selected cases were set to be analysed. To

Table 4.2 Mechanisms and locations of the selected accidents from the NCIS (accidents with fewer than five incidents were deleted from the table due to identifiability)

	Victoria	*Queensland*	*Western Australia*	*Tasmania*
Fall from height	16	10	7	-
Contact with electricity	8	10	0	0
Being hit by falling objects	0	-	5	0
Being hit by moving objects	5	5	0	0
Trapped between or in equipment	-	9	-	0
Vehicle collision	0	-	-	0
Other	8	-	-	0

elaborate on the content analysis, codes discovered in seven cases that involved the seven accident mechanisms are described herein.

1) Case No. 1 (other type accident mechanism)

    i.   The sub-contractor employed an untrained person for a hazardous job (coded as safety culture and worker capability).

    ii.   At the time of the incident, the recorded temperature was 42.1 °C and the employee was working on the roof under direct sun, which considerably increased the intensity of the hot temperature (coded as working environment).

    iii.   A colleague of the deceased noticed the deceased was suffering from serious heat stress (coded as worker health).

    iv.   The company did not have a risk assessment for heat stress conditions (coded as risk management).

    v.   Workers had long working hours (coded as scheduling).

    vi.   The project manager did not call work off when he/she noticed the temperature reached 38 °C (coded as project management).

    vii.   The deceased did not have any training regarding the performed job (coded as worker capability).

    viii.   There was no site supervisor on site on that particular day (coded as immediate supervision).

2) Case No. 2 (trapped in or between equipment accident mechanism)

    i.   There was no communication, and the equipment operator thought the deceased was not in the vicinity of the equipment (coded as communication).

    ii.   The deceased was standing on the blind side of the equipment. The normal practice was to leave the area of the truck driver during the refuelling (unsafe action and behaviour).

    iii.   The deceased and some other equipment operators did not receive any training or instruction on the refuelling procedure (coded as worker capability).

    iv.   Although there was a safe work method statement on the site, it was not provided to everyone, including the deceased (coded as construction process).

3) Case No. 7 (contact-with-electricity accident mechanism)

    i.   Investigations found there was a steel bar wedged into a floor bearer under the floor, right next to an electrical conduit where the deceased was electrocuted (coded as permanent work design).

    ii.   The deceased was not wearing any footwear at the time of the incident (coded as unsafe action).

    iii.   No site supervisor was available on the site (coded as immediate supervision).

    iv.   There was no proper risk assessment or work method statement for working under the floor (coded as risk management and construction process).

4) Case No. 10 (being hit by moving object accident mechanism)

i.   The company did not conduct appropriate risk assessment of the equipment (risk management).
ii. The equipment that the deceased was standing on did not have an emergency stop mechanism (equipment condition).
iii. The deceased stood on equipment that they did not have expertise in using (coded as unsafe action and worker capability).
iv. The deceased was working alone, and there was no supervisor available (coded as immediate supervision).

5) Case No. 23 (fall-from-height accident mechanism)

i.   The main contractor allowed for the use of equipment (elevated work platform boom lift) that was not specifically related to the site and was not suitable for the activity (coded as project management and immediate supervision).
ii. The sub-contractor initiated the use of the equipment without submitting a risk assessment document or work method statement (coded as construction process and risk management).
iii. The equipment was on an area of uncompacted ground that was sloping (coded as site condition).
iv. The deceased was not wearing safety PPE at the time of the incident (coded as unsafe action and behaviour).
v.   Expert investigators noted that the selected equipment was not suitable for the site condition (coded as suitable equipment).

6) Case No. 74 (being hit by falling objects accident mechanism)

i.   There appeared to be no proper inspection of the lifting arrangement by the crane operator or the dogman (coded as immediate supervision).
ii. The laminated lifting beams snapped and the shutter fell vertically to the ground, onto the plywood side, and onto the deceased's lower body (coded as a material condition).
iii. The deceased was the site supervisor that designed the method of connection and extension of the header beams to the existing shutter. The deceased did not have a certificate for such a design practice (coded as the construction process, project management and worker capability).
iv. The screws between the brace and the uprights were inadequate and did not comply with the standard requirements for screwed connections (AS3610) (coded as equipment suitability)
v.   The deceased was standing at an unsafe proximity when the incident occurred (coded as unsafe action and behaviour)

7) Case No. 100 (vehicle collision accident mechanism)

i.   Investigations revealed that the excavator had rolled due to being too close to the edge of a dam with a steep incline on the southern side (coded as construction process, immediate supervision).

ii. The equipment did not have a rollover protection structure and the absence of this may have contributed to the deceased's death (equipment condition).
iii. Investigations revealed that the deceased had been working for many hours over many days on the project (coded as scheduling) and that fatigue (coded as worker health/fatigue) and operator error (coded as worker action) may also have played a part in the incident.

To elaborate on the most frequently discovered factors, a summary of the 487 identified factors is provided in Table 4.3. These factors are grouped based on the accident mechanism categories of the NCIS (111). Overall, the highest number of causations were attributed to immediate supervision (60 cases) and workers' actions and behaviour (58 cases), followed by risk management (54 cases), construction process (48 cases) and permanent work design (45 cases) from the originating influences group. These figures also show that, in total, the originating influences had the highest contribution to risk (42.5%), followed by immediate accident circumstances (31.8%) and shaping factors (25.7%).

However, the figures varied when the accident mechanisms were compared. In fall-from-height accidents, immediate supervision, permanent work design, workers' actions and behaviours and construction processes were the highest contributing factors. In contact-with-electricity accidents, permanent work design, construction process and risk management were found more frequently. Workers' actions and behaviours were at the top of the contributing factors list for being hit by moving objects and vehicle collision accidents, while immediate supervision was the most frequent factor in being hit by falling objects and trapped between or in equipment accidents. Risk management and permanent work design had the highest scores in the other categories.

4.5   Accident Causation Significance Analysis

4.5.1   *Standardised Degree Centrality Analysis*

Social network analysis (SNA) was initially introduced to investigate the interactions of members within a group. Moreno (119) defined SNA as "a quantitative analytic tool used to study the exchange of resources among different groups" (p. 17). Later, Haythornthwaite (120) defined it as "an approach and set of techniques used to study the exchange of resources among actors" (p. 323). Despite the definitions put forward by several researchers, Wasserman and Faust (121) noted that the main advantage of SNA is its ability to provide a researcher with an analytical tool to investigate the pattern of network connections among its actors and to visualise and quantify some important metrics. According to Freeman (122), these metrics include network density, actor centrality and betweenness.

The ConAC model presents a complex network of causations that can lead to accidents. Considering accident causations as the actors in the ConAC model allows for analysis of the network characteristics of accident causations. Several researchers have attempted to employ SNA methods to understand the pattern of

Table 4.3 Summary of the content analysis results according to accident mechanisms

	Fall from height	Contact with electricity	Being hit by moving objects	Being hit by falling objects	Trapped between or in equipment	Vehicle collision	Other	Total
Worker actions and behaviours	19	11	8	4	9	3	4	58
Worker capabilities	8	7	2	2	7	0	3	29
Communication	2	2	1	0	4	0	4	13
Immediate supervision	24	7	6	7	10	1	5	60
Worker health/fatigue	5	0	0	0	1	1	6	13
Site conditions	2	0	3	1	3	1	2	12
Site constraints, site layout/space	0	4	0	1	2	0	0	7
Local hazards	0	3	0	0	0	0	2	5
Working environment	1	2	1	1	0	0	4	9
Work scheduling	0	2	0	0	0	1	1	4
Housekeeping	0	1	0	0	0	0	0	1
Suitability of materials	1	1	0	1	0	0	1	4
Usability of materials	1	3	0	0	0	0	0	4
Condition of materials	2	6	0	2	0	0	2	12
Suitability of equipment	11	2	0	2	3	0	1	19
Usability of equipment	2	1	1	0	2	0	1	7
Condition of equipment	8	0	3	3	4	2	3	23
Permanent (and temporary) work design	20	7	3	1	4	2	8	45
Project management	9	4	4	2	3	0	3	25
Construction processes	15	11	4	3	8	2	5	48
Safety culture	14	6	3	4	3	1	4	35
Risk management	14	12	7	4	7	1	9	54
Total number of cases	34	18	10	8	15	4	16	487

construction accidents (123, 124). In SNA, the degree of centrality is a primary metric to identify the most connected actors in a network. This measure implies the relative extent of an actor's connections with the remainder of a network (125). A construction accident involves a chain of actors (causations) that lead to the occurrence of an accident, as described in the ConAC model (30). Finding the central factors of accidents can reveal the most effective areas in which to intervene and reduce the likelihood of accidents. Therefore, in this step, the centralities of the causations in the accident networks were analysed. This was achieved by calculating the standardised degree centralities (C_D) of the identified causation in each case using Equation 3.1 (124, 126).

$$C_D \text{ (actor x)} = cD \text{ (actor x)} / (g - 1) \qquad (3.1)$$

where C_D (actor x) is the sum of relationships that actor x has and g is the total number of actors in the network. Relationships among the actors were extracted from the ConAC model and presented in Figure 4.2, where the actors of the originating influences layer are directly connected to the actors of the shaping factors layer. Also, depending on the type of actor in the shaping factors layer (worker-, workplace-, material- or equipment-related), the actors of the shaping factors layer are connected to the actors of the immediate accident circumstances layer. Based on these relationships, networks for each of the accident cases were developed using the findings of the thematic content analysis, and standardised centralities were calculated for each of the actors.

To this end, mean standardised centralities were calculated for the actors in each category of the accident mechanisms. de Brito, do Socorro da Silva (127) suggest categorising actors into a network into four groups of very low, low, medium and high

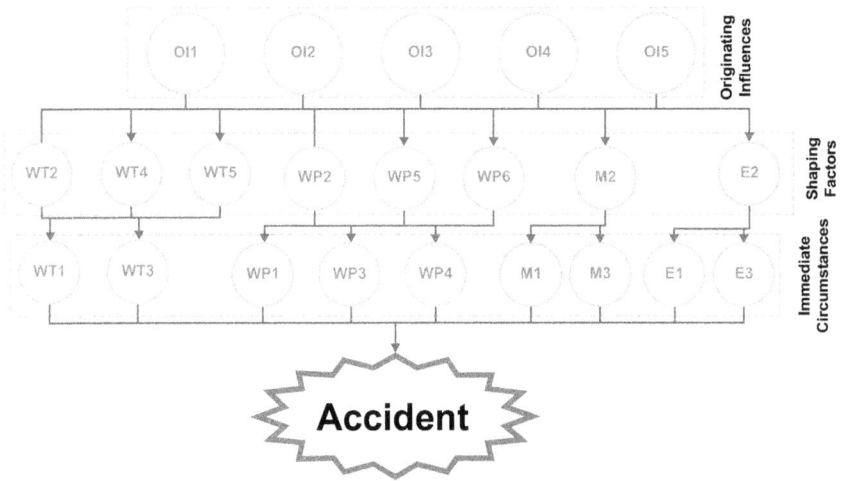

Figure 4.2 Relationship of actors in the ConAC accident model

centrality. This can be achieved by classifying the centrality indices as defining 25% of entries in each range. High and medium central causations are considered critical causations in accident cases and were used to address Objective 1 of this study.

4.5.2   Accident Causation Significance Analysis

As illustrated in Table 4.4, in the fall-from-height accidents, highly central factors were immediate supervision (WT4) and temporary and permanent work design (OI1), while medium central factors were construction process (OI3) and workers' actions and behaviours (WT1). Figure 4.3 presents C_D values of accident causations for fall-from-height accidents and shows that most of the ConAC factors were involved in the accidents, except the workplace factors of site constraints (WP2), local hazards (WP3), work scheduling (WP5) and housekeeping (WP6). The lowest C_D value found in this category was for WP4 (0.007), and the highest value was for WT4 (0.459).

Table 4.4 Ranges of standardised degree centrality (C_D) of the causations of fall-from-height accidents

Category	Ranges of standardised degree centrality based on 25% intervals from the lowest to the highest margins	Accident causations
Very low	0.007 to 0.120	WT3, M1, M3, E3, WT5, M2, E2, WP1, WP4
Low	0.121 to 0.223	E3, WT2, OI2, OI4, OI5
Medium	0.224 to 0.346	WT1, OI3
High	0.347 to 0.459	WT4, OI1

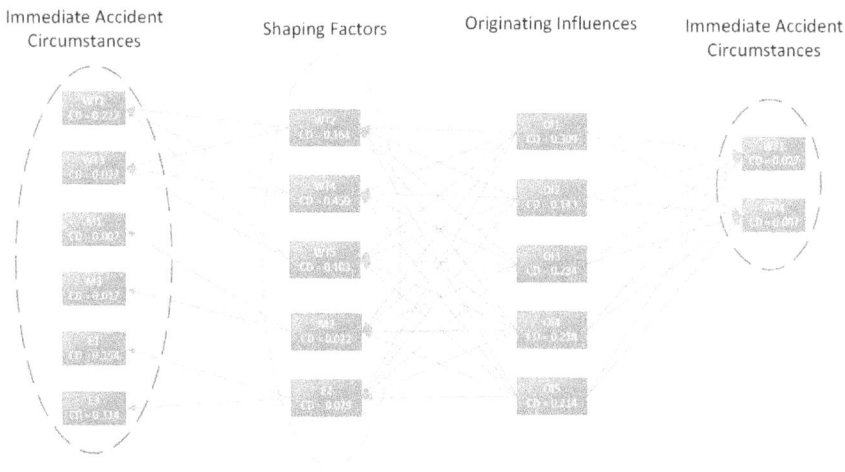

Figure 4.3 Standardised degree centralities (C_D) of the fall-from-height accident causations

Among the contact-with-electricity accident causations (Table 4.5), risk management (OI5) and construction process (OI3) were highly central. Workers' capabilities (WT2), immediate supervision (WT4), workers' actions and behaviours (WT1) and temporary and permanent work design (OI1) were moderately central. As in the case of fall-from-height accidents, originating influences had the highest overall centrality in the contact with electricity category (Figure 4.4). The highest

Table 4.5 Ranges of standardised degree centrality (C_D) of contact-with-electricity accident causations

Category	Ranges of standardised degree centrality based on 25% intervals from the lowest to the highest margins	Accident causations
Very low	0.014 to 0.100	WT3, WP3, WP4, M1, E1, WP2, WP5, WP6, M2, E2, OI2
Low	0.101 to 0.186	M3, OI4
Medium	0.187 to 0.272	WT1, WT2, WT4, OI1
High	0.273 to 0.358	OI5, OI3

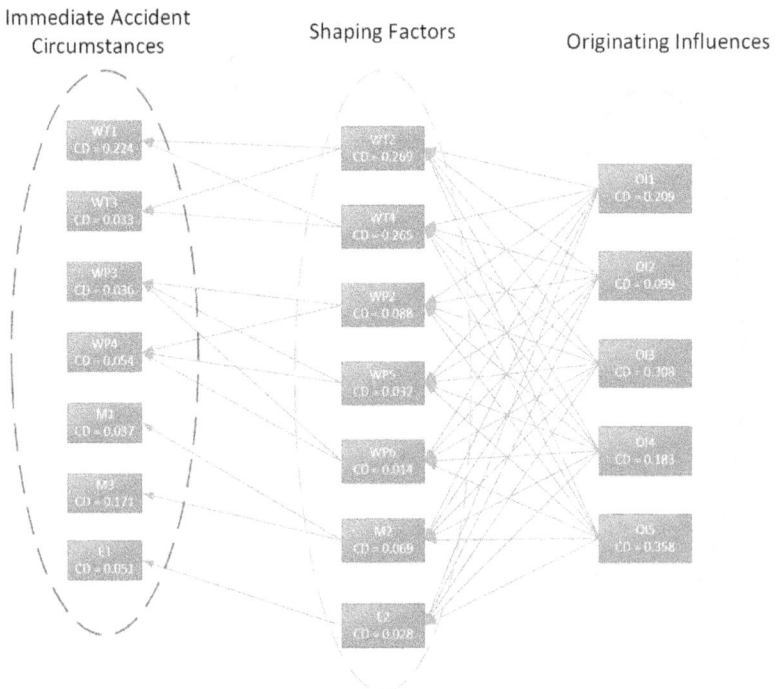

Figure 4.4 Standardised degree centralities (C_D) of contact-with-electricity accident causations

C_D value in this category was for OI5 (0.358), and the lowest value was for WP6 (0.014).

Immediate supervision (WT4) was found to be the most central factor in being hit by falling objects accidents, with a 0.544 standardised centrality value, followed by risk management (OI5) with a medium standardised centrality value of 0.377 (Table 4.6). Figure 4.5 shows that the overall centralities of the upper layers of the ConAC model were higher than those of the lower layers, and the likelihood of accidents could be reduced significantly in those levels. The highest C_D value in this category was for WT4 (0.554) and the lowest value was for WP6 (0.014).

Table 4.7 demonstrates the centralities of the causations of being hit by moving objects accidents. These accidents were most influenced by three major causations: Immediate supervision (WT4), risk management (OI5) and workers' actions and behaviours (WT1), which were categorised as highly central. There was no

Table 4.6 Ranges of standardised degree centrality (C_D) of being hit by falling objects accident causations

Category	Ranges of standardised degree centrality based on 25% intervals from the lowest to the highest margins	Accident causations
Very low	0.014 to 0.146	WP1, WP4, WT2, WP2, WP5, OI1, OI2, M1, M3, E1
Low	0.147 to 0.278	WT1, OI3, OI4, E3
Medium	0.279 to 0.411	OI5
High	0.412 to 0.544	WT4

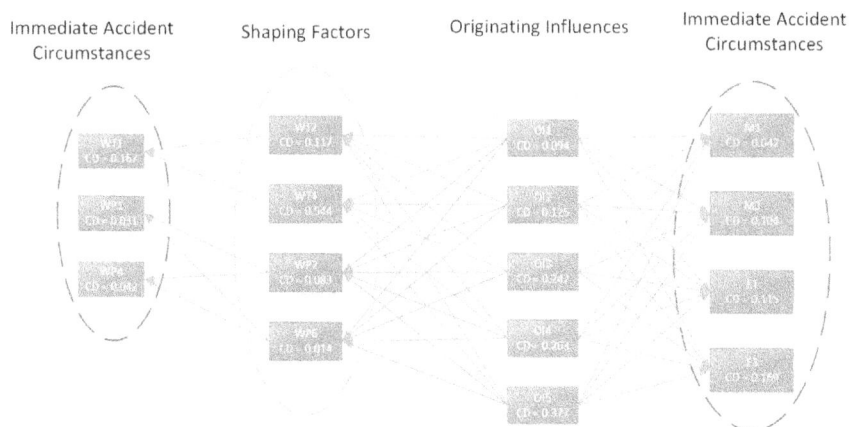

Figure 4.5 Standardised degree centralities (C_D) of being hit by falling objects accident causations

causation in the medium centrality category. Figure 4.6 shows that the number of shaping factors in this category was less than that for fall from height and contact-with-electricity accidents. The highest C_D value in this category was for WT4 (0.405), and the lowest value was for E2 (0.06).

Among the trapped between or in equipment accident causations, immediate supervision (WT4) and workers' capabilities (WT2) were highly central (Table 4.8). In addition, construction process (OI3), risk management (OI5) and workers' actions and behaviours (WT1) had medium centrality in this category. Figure 4.7 shows that most causations were related to applying a safe working process. The highest C_D value in this category was for WT4 (0.396), and the lowest value was for WT5 (0.024). Material-related causations did not contribute to any of the accidents in this category.

Table 4.7 Ranges of standardised degree centrality (C_D) of being hit by moving objects accident causations

Category	Ranges of standardised degree centrality based on 25% intervals from the lowest to the highest margins	Accident causations
Very low	0.060 to 0.144	WT3, E3, WT2, E2, WP4
Low	0.145 to 0.229	OI1, OI2, OI3, OI4, WP1
Medium	0.230 to 0.313	-
High	0.314 to 0.398	WT1, WT4, OI5

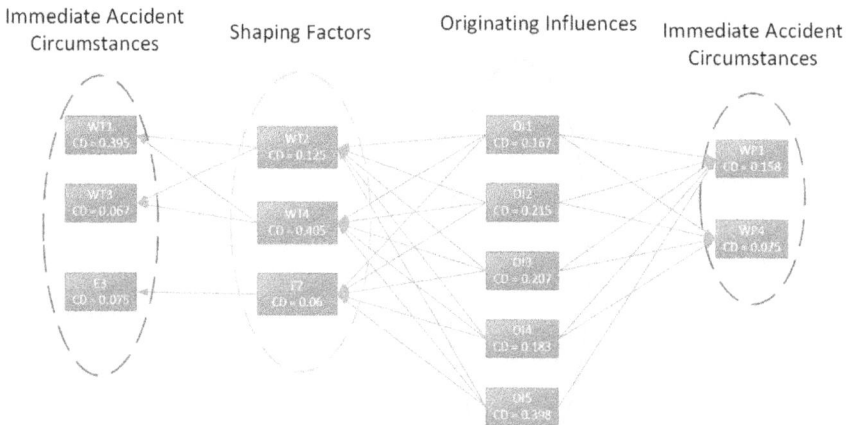

Figure 4.6 Standardised degree centralities (C_D) of being hit by moving objects accident causations

Table 4.8 Ranges of standardised degree centrality (C_D) of trapped between or in equipment accident causations

Category	Ranges of standardised degree centrality based on 25% intervals from the lowest to the highest margins	Accident causations
Very low	0.024 to 0.117	WP1, E1, E3, WT5, WP2, E2, OI2, OI4
Low	0.118 to 0.210	WT3, OI1
Medium	0.211 to 0.303	WT1, OI3, OI5
High	0.304 to 0.396	WT2, WT4

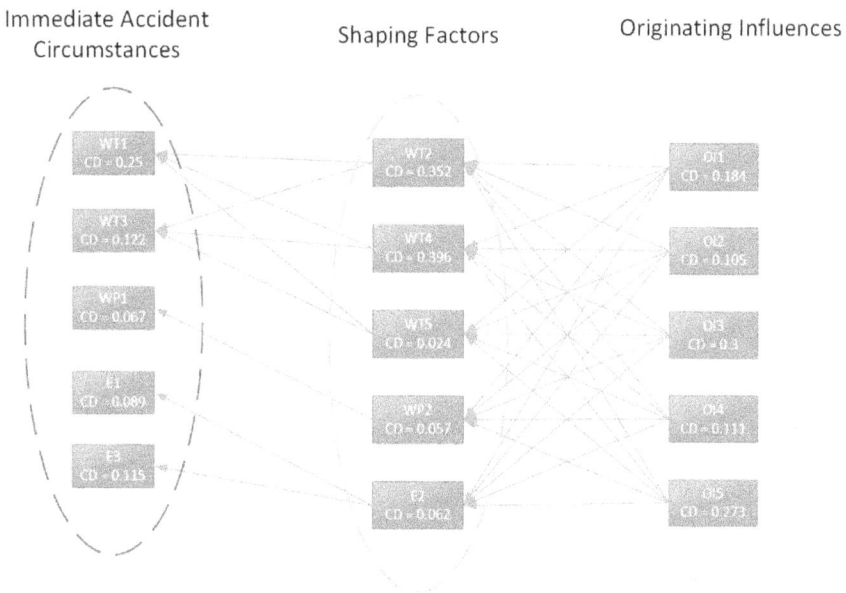

Figure 4.7 Standardised degree centralities (C_D) of trapped between or in equipment accident causations

Workers' actions and behaviours (WT1) was the most highly central actor attributed to the vehicle collision accidents, followed by medium central factors of permanent work design (OI1), equipment condition (E3) and construction process (OI3) (Table 4.9). Equipment-related and workplace-related shaping factors, along with material-related factors, did not contribute to this type of accident (Figure 4.8). The highest C_D value in this category was for WT1 (0.475), and the lowest value was for OI5 (0.083).

Other accident types were defined as those where the mechanism of the accidents was not among the listed mechanisms. In these types of accidents, permanent

Table 4.9 Ranges of standardised degree centrality (C_D) of vehicle collision accident causations

Category	Ranges of standardised degree centrality based on 25% intervals from the lowest to the highest margins	Accident causations
Very low	0.083 to 0.181	WT4, WT5, OI5, WP1
Low	0.182 to 0.279	OI4
Medium	0.280 to 0.377	OI1, OI3, E3
High	0.378 to 0.475	WT1

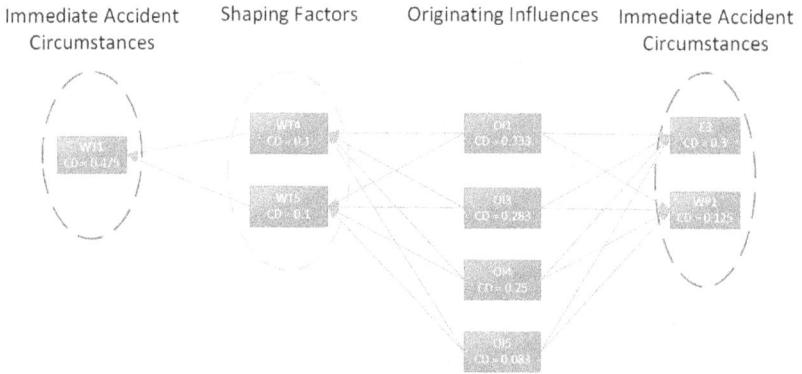

Figure 4.8 Standardised degree centralities (C_D) of vehicle collision accident causations

Table 4.10 Ranges of standardised degree centrality (C_D) of other accident causations

Category	Ranges of standardised degree centrality based on 25% intervals from the lowest to the highest margins	Accident causations
Very low	0.014 to 0.080	WP1, WP3, WP4, E1, WP5, WP6, E2, M1
Low	0.081 to 0.147	WT1, WP4, WT3, E3, WT2, OI2, M3
Medium	0.148 to 0.213	WT4, OI3
High	0.214 to 0.280	WT5, OI1, OI5

work design (OI1), risk management (OI5) and workers' health and fatigue (WT5) were ranked as highly central. In addition, safety culture (OI4) and immediate supervision were ranked as moderately central in this category (Table 4.10). Figure 4.9 illustrates that most of the ConAC factors were involved in accidents of this

Immediate Accident
Circumstances

Shaping Factors

Originating Influences

Immediate Accident
Circumstances

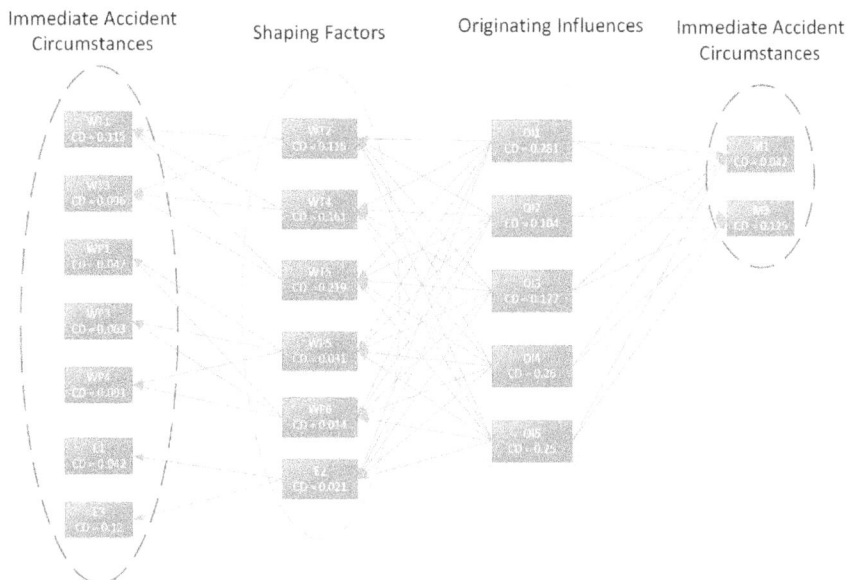

Figure 4.9 Standardised degree centralities (C_D) of other type accident causations

type, except site constraint (WP2) and material usability (E2). The highest C_D value in this category was for OI1 (0.281), and the lowest value was for WP6 (0.014).

4.6 Summary

A description of accident theories within and outside of the construction WHS literature was provided in this chapter. A detailed discussion of the ConAC model was provided as the most appropriate model to handle construction accidents. Using this model, a total of 105 accident cases were analysed using the thematic content analysis method. The standardised degree centrality of accident causations was then calculated for different accident categories, and their significance was evaluated. To summarise the findings of accident analysis, the high and medium central actors in the seven accident mechanisms were: Workers' actions and behaviours (WT1), worker capabilities (WT2), immediate supervision (WT4), worker health and fatigue (WT5), permanent (and temporary) work design (OI1), construction process (OI3) and risk management (OI5).

References

1. Katsakiori P, Sakellaropoulos G, Manatakis E. Towards an evaluation of accident investigation methods in terms of their alignment with accident causation models. *Safety Science*. 2009;47(7):1007–15.

2. Hosseinian SS, Torghabeh ZJ. Major theories of construction accident causation models: a literature review. *International Journal of Advances in Engineering & Technology*. 2012;4(2):53.
3. Bird FE, Loftus RG. *Loss Control Management*. Loganville, GA: Institute Press; 1976.
4. Svenson O. The accident evolution and barrier function (AEB) model applied to incident analysis in the processing industries. *Risk Analysis*. 1991;11(3):499–507.
5. Hollnagel E. Time and time again. *Theoretical Issues in Ergonomics Science*. 2002;3(2):143–58.
6. Goh YM, Brown H, Spickett J. Applying systems thinking concepts in the analysis of major incidents and safety culture. *Safety Science*. 2010;48(3):302–9.
7. Gordon R, Flin R, Mearns K. Designing and evaluating a human factors investigation tool (HFIT) for accident analysis. *Safety Science*. 2005;43(3):147–71.
8. Endsley MR, Garland DJ. *Situation Awareness Analysis and Measurement*. New York: CRC Press; 2000.
9. Behm M, Schneller A. Application of the loughborough construction accident causation model: a framework for organizational learning. *Construction Management and Economics*. 2013;31(6):580–95.
10. Gibb A, Lingard H, Behm M, Cooke T. Construction accident causality: learning from different countries and differing consequences. *Construction Management and Economics*. 2014;32(5):446–59.
11. Reason J. Human error: models and management. *BMJ: British Medical Journal*. 2000;320(7237):768–70.
12. Rasmussen J, Vicente KJ. Coping with human errors through system design: implications for ecological interface design. *International Journal of Man-Machine Studies*. 1989;31(5):517–34.
13. Culver C, Marshall M, Connolly C. Analysis of construction accidents: the workers' compensation database. *Professional Safety*. 1993;38(3):22.
14. Hinze J, Russell DB. Analysis of fatalities recorded by OSHA. *Journal of Construction Engineering and Management*. 1995;121(2):209–14.
15. Hunting KL, Nessel-Stephens L, Sanford SM, Shesser R, Welch LS. Surveillance of construction worker injuries through an urban emergency department. *Journal of Occupational and Environmental Medicine*. 1994;36(3):356–64.
16. Kisner SM, Fosbroke DE. Injury hazards in the construction industry. *Journal of Occupational and Environmental Medicine*. 1994;36(2):137–43.
17. Snashall D. Safety and health in the construction industry. *BMJ: British Medical Journal*. 1990;301(6752):563.
18. Daniels C, Marlow P. Literature review on the reporting of workplace injury trends. *Health and Safety Laboratory*. 2005;36.
19. Bomel L. Improving health and safety in construction: phase 1: data collection, review and structuring. *Contract Research Report*. 2001;387.
20. Gyi DE, Gibb AG, Haslam RA. The quality of accident and health data in the construction industry: interviews with senior managers. *Construction Management & Economics*. 1999;17(2):197–204.
21. Hinze J, editor. *The Construction Safety Record Since 1971*. Civil Engineers Influencing Public Policy: ASCE; 1996.
22. Abdelhamid TS, Everett JG. Identifying root causes of construction accidents. *Journal of Construction Engineering and Management*. 2000;126(1):52–60.
23. Whittington C, Livingston A, Lucas D. *Research into Management, Organisational and Human Factors in the Construction Industry*. Sudbury, Suffolk: Health and Safety Executive; 1992.

24. Suraji A, Duff AR, Peckitt SJ. Development of causal model of construction accident causation. *Journal of Construction Engineering and Management*. 2001;127(4):337–44.
25. Priemus H, Ale B. Construction safety: an analysis of systems failure: the case of the multifunctional Bos & Lommerplein estate, Amsterdam. *Safety Science*. 2010;48(2):111–22.
26. Manu P, Ankrah N, Proverbs D, Suresh S. An approach for determining the extent of contribution of construction project features to accident causation. *Safety Science*. 2010;48(6):687–92.
27. Mitropoulos P, Abdelhamid TS, Howell GA. Systems model of construction accident causation. *Journal of Construction Engineering and Management*. 2005;131(7):816–25.
28. Hale A, Walker D, Walters N, Bolt H. Developing the understanding of underlying causes of construction fatal accidents. *Safety Science*. 2012;50(10):2020–7.
29. Garrett JW, Teizer J. Human factors analysis classification system relating to human error awareness taxonomy in construction safety. *Journal of Construction Engineering and Management*. 2009;135(8):754–63.
30. Gibb AG, Haslam R, Gyi DE, Hide S, Duff R. What causes accidents? *Proceedings of the Institution of Civil Engineers*. 2006;159(6):46–50.
31. Phua F. Does the built-environment industry attract risk-taking individuals? *Construction Management and Economics*. 2016:1–11.
32. Alsamadani R, Hallowell MR, Javernick-Will A, Cabello J. Relationships among language proficiency, communication patterns, and safety performance in small work crews in the United States. *Journal of Construction Engineering and Management*. 2013;139(9):1125–34.
33. Center for Construction Research and Training. *The Construction Chart Book* (5th ed.). Silver Spring, Maryland: The U.S. Construction Industry and Its Workers; 2013.
34. Arndt V, Rothenbacher D, Daniel U, Zschenderlein B, Schuberth S, Brenner H. All-cause and cause specific mortality in a cohort of 20 000 construction workers; results from a 10 year follow up. *Occupational and Environmental Medicine*. 2004;61(5):419–25.
35. Comu S, Unsal HI, Taylor JE. Dual impact of cultural and linguistic diversity on project network performance. *Journal of Management in Engineering*. 2011;27(3):179–87.
36. Vecchio-Sadus AM. Enhancing safety culture through effective communication. *Safety Science Monitor*. 2007;11(3):1–10.
37. Perlman A, Sacks R, Barak R. Hazard recognition and risk perception in construction. *Safety Science*. 2014;64:22–31.
38. Haslam RA, Hide SA, Gibb AGF, Gyi DE, Pavitt T, Atkinson S, et al. Contributing factors in construction accidents. *Applied Ergonomics*. 2005;36(4):401–15.
39. Albert A, Hallowell MR, Kleiner B, Chen A, Golparvar-Fard M. Enhancing construction hazard recognition with high-fidelity augmented virtuality. *Journal of Construction Engineering and Management*. 2014;140(7):04014024.
40. Fleming P. *Authenticity and the Cultural Politics of Work: New Forms of Informal Control*. Oxford: Oxford University Press; 2009.
41. Jeelani I, Albert A, Gambatese JA. Why do construction hazards remain unrecognized at the work interface? *Journal of Construction Engineering and Management*. 2017;143(5):04016128.
42. Hinze J. *Construction Safety*. New York: Prentice Hall; 2006.
43. Fam IM, Nikoomaram H, Soltanian A. Comparative analysis of creative and classic training methods in health, safety and environment (HSE) participation improvement. *Journal of Loss Prevention in the Process Industries*. 2012;25(2):250–3.
44. Hallowell MR, van Boven L, Tixier AJP, Albert A, Kleiner BM. Psychological antecedents of risk-taking behavior in construction. *Journal of Construction Engineering and Management*. 2014;140(11):4014052.

45. Namian M, Albert A, Behm M, Zuluaga CM. Role of safety training: impact on hazard recognition and safety risk perception. *Journal of Construction Engineering and Management*. 2016;142(12):4016073.
46. Li H, Chan G, Skitmore M. Visualizing safety assessment by integrating the use of game technology. *Automation in Construction*. 2012;22:498–505.
47. Mohammadfam I, Ghasemi F, Kalatpour O, Moghimbeigi A. Constructing a Bayesian network model for improving safety behavior of employees at workplaces. *Applied Ergonomics*. 2017;58:35–47.
48. Toole TM. Construction site safety roles. *Journal of Construction Engineering and Management*. 2002;128(3):203–10.
49. Yi W, Chan APC, Wang X, Wang J. Development of an early-warning system for site work in hot and humid environments: a case study. *Automation in Construction*. 2016;62:101–13.
50. Chan APC, Yi W, Wong DP, Yam MCH, Chan DWM. Determining an optimal recovery time for construction rebar workers after working to exhaustion in a hot and humid environment. *Building and Environment*. 2012;58:163–71.
51. Lucas RAI, Epstein Y, Kjellstrom T. Excessive occupational heat exposure: a significant ergonomic challenge and health risk for current and future workers. *Extreme Physiology & Medicine*. 2014;3(1):14.
52. Yi W, Chan APC. Critical review of labor productivity research in construction journals. *Journal of Management in Engineering*. 2014;30(2):214–25.
53. Taylor NAS. Challenges to temperature regulation when working in hot environments. *Industrial Health*. 2006;44(3):331–44.
54. Horie S. Prevention of heat stress disorders in the workplace. *JMAJ*. 2013;56(3):186–92.
55. Petitti DB, Harlan SL, Chowell-Puente G, Ruddell D. Occupation and environmental heat-associated deaths in maricopa county, Arizona: a case-control study. *PLoS ONE*. 2013;8(5):e62596.
56. Joubert D, Thomsen J, Harrison O. Safety in the heat: a comprehensive program for prevention of heat illness among workers in Abu Dhabi, United Arab Emirates. *American Journal of Public Health*. 2011;101(3):395–8.
57. Huang C, Wong CK. Optimisation of site layout planning for multiple construction stages with safety considerations and requirements. *Automation in Construction*. 2015;53:58–68.
58. Payne E, Van Valkenburgh NS, Van Valkenburgh GL. Modular mobile safety structure for containment and handling of hazardous materials. *Google Patents*. United States patent US 5,735,639; 1998.
59. LaDou J, Landrigan P, Baila JC, Foa V, Frank A. A call for an international ban on asbestos. *Canadian Medical Association Journal*. 2001;164(4):489.
60. Bartrip P, editor. Asbestos and health in twentieth century Britain. Motives and outcomes. *14th Interanational Economic History Congress (IEHC)*. Helsinki: IEHC; 13 July 2006: 21–5.
61. Chong D, Wang Y, Guo H, Lu Y. Volatile organic compounds generated in asphalt pavement construction and their health effects on workers. *Journal of Construction Engineering and Management*. 2014;140(2):04013051.
62. Holt GD, Edwards DJ. Machinery transportation management: case study of "plant-trailer" H&S incidents. *Built Environment Project and Asset Management*. 2014;4(3):264–80.
63. Edwards DJ, Love PED. A case study of machinery maintenance protocols and procedures within the UK utilities sector. *Accident Analysis & Prevention*. 2016;93:319–29.

64. Holt GD. Opposing influences on construction plant and machinery health and safety innovations. *Construction Innovation*. 2016;16(3):390–414.

65. Brown KA, Willis PG, Prussia GE. Predicting safe employee behavior in the steel industry: development and test of a sociotechnical model. *Journal of Operations Management*. 2000;18(4):445–65.

66. MacCrimmon KR, Wehrung D, Stanbury WT. *Taking Risks*. Simon and Schuster; 1988.

67. March JG, Shapira Z. Variable risk preferences and the focus of attention. *Psychological Review*. 1992;99(1):172.

68. DeJoy DM. Theoretical models of health behavior and workplace self-protective behavior. *Journal of Safety Research*. 1996;27(2):61–72.

69. Mischke C, Verbeek JH, Job J, Morata TC, Alvesalo-Kuusi A, Neuvonen K, et al. Occupational safety and health enforcement tools for preventing occupational diseases and injuries. *Cochrane Database of Systematic Reviews*. 2013;8.

70. Tompa E, Trevithick S, McLeod C. Systematic review of the prevention incentives of insurance and regulatory mechanisms for occupational health and safety. *Scandinavian Journal of Work, Environment & Health*. 2007;33(2):85–95.

71. Grandjean E. Fatigue in industry. *British Journal of Industrial Medicine*. 1979;36(3):175.

72. Uher T, Zantis AS. *Programming and Scheduling Techniques*. Routledge; 2012.

73. Bakry I, Moselhi O, Zayed T, editors. Fuzzy dynamic programming for optimized scheduling of repetitive construction projects. *2013 Joint IFSA World Congress and NAFIPS Annual Meeting (IFSA/NAFIPS)*. IEEE; 24–28 June 2013: 1172–6.

74. Lucko G. Integrating efficient resource optimization and linear schedule analysis with singularity functions. *Journal of Construction Engineering and Management*. 2011;137(1):45–55.

75. Hegazy T, Kamarah E. Efficient repetitive scheduling for high-rise construction. *Journal of Construction Engineering and Management*. 2008;134(4):253–64.

76. Esmaeili B, Hallowell M. Integration of safety risk data with highway construction schedules. *Construction Management and Economics*. 2013;31(6):528–41.

77. Goh YM, Askar Ali MJ. A hybrid simulation approach for integrating safety behavior into construction planning: an earthmoving case study. *Accident Analysis & Prevention*. 2016;93:310–18.

78. Kassem M, Dawood N, Chavada R. Construction workspace management within an industry foundation class-compliant 4D tool. *Automation in Construction*. 2015;52:42–58.

79. Kartam NA. Integrating safety and health performance into construction CPM. *Journal of Construction Engineering and Management*. 1997;123(2):121–6.

80. Hinze J, Nelson J, Evans R, editors. Software integration of safety in construction schedules. *Proceedings of the 4th Triennial International Conference, Rethinking and Revitalizing Construction Safety*. Health, Environment and Quality, Port Elizabeth, South, Africa; 2005.

81. Yi K-J, Langford D. Scheduling-based risk estimation and safety planning for construction projects. *Journal of Construction Engineering and Management*. 2006;132(6):626–35.

82. Wang W-C, Liu J-J, Chou S-C. Simulation-based safety evaluation model integrated with network schedule. *Automation in Construction*. 2006;15(3):341–54.

83. Bluff E. Safety in machinery design and construction: performance for substantive safety outcomes. *Safety Science*. 2014;66:27–35.

84. Brauer RL. *Safety and Health for Engineers*. Hoboken, New Jersey: John Wiley & Sons; 2016.

85. Szymberski R. Construction project safety planning. *Tappi Journal (USA)*. 1997;80(11):69–74.

86. Behm M. Linking construction fatalities to the design for construction safety concept. *Safety Science*. 2005;43(8):589–611.
87. Gibb A, Haslam R, Hide S, Gyi D. The role of design in accident causality. *Designing for Safety and Health in Construction*. 2004:11–21.
88. Safe Work Australia. *Design Issues in Work-Related Serious Injuries 2009*. Available from: www.safeworkaustralia.gov.au/sites/SWA/about/Publications/Documents/286/DesignIssues_WorkRelatedSeriousInjuries_2005_PDF.pdf.
89. Breslin P. Improving OHS standards in the building and construction industry through safe design. *Journal of Occupational Health and Safety, Australia and New Zealand*. 2007;23(1):89.
90. Tymvios N, Gambatese J, Sillars D, editors. Designer, contractor, and owner views on the topic of design for construction worker safety. *Construction Research Congress 2012: Construction Challenges in a Flat World*. West Lafayette, IN; 21–23 May 2012: 341–55.
91. Hinze J, Wiegand F. Role of designers in construction worker safety. *Journal of Construction Engineering and Management*. 1992;118(4):677–84.
92. Gambatese JA, Behm M, Hinze JW. Viability of designing for construction worker safety. *Journal of Construction Engineering and Management*. 2005;131(9):1029–36.
93. Health and Safety Executive. *Construction (Design and Management) Regulation UK: Health and Safety Executive*; 2016. Available from: www.hse.gov.uk/.
94. Wright M, Berman G. *The Case for CDM: Better Safer Design: A Pilot Study*. Suffolk: HSE Books; 2003.
95. Workcover A. *Construction Hazard Assessment Implication Review – CHAIR*; 2001. Available from: https://www.safedesignaustralia.com.au/wp-content/uploads/2018/10/CHAIR_Safety_in_Design_Tool_WorkCoverNSW.pdf.
96. Pinion C, Brewer S, Douphrate D, Whitehead L, DelliFraine J, Taylor WC, et al. The impact of job control on employee perception of management commitment to safety. *Safety Science*. 2017;93:70–5.
97. Yule S. Senior management influence on safety in the UK and US energy sectors. Doctoral dissertation. Aberdeen: University of Aberdeen; 2003.
98. Zohar D. Thirty years of safety climate research: reflections and future directions. *Accident Analysis & Prevention*. 2010;42(5):1517–22.
99. Cooke T, Lingard H, editors. A retrospective analysis of work-related deaths in the Australian construction industry. *ARCOM Twenty-seventh Annual Conference*. Association of Researchers in Construction Management (ARCOM); 2011.
100. Lingard H, Cooke T, Gharaie E. A case study analysis of fatal incidents involving excavators in the Australian construction industry. *Engineering, Construction and Architectural Management*. 2013;20(5):488–504.
101. Safe Work Australia. *Work-related Injuries and Fatalities in Construction, Australia, 2003 to 2013*. Australia; 2015.
102. Cameron I, Hare B, Davies R. Fatal and major construction accidents: a comparison between Scotland and the rest of Great Britain. *Safety Science*. 2008;46(4):692–708.
103. Golizadeh H, Hon CKH, Drogemuller R, Hosseini MR. Digital engineering potential in addressing causes of construction accidents. *Automation in Construction*. 2018;95:284–95.
104. Tixier AJP, Hallowell MR, Rajagopalan B, Bowman D. Application of machine learning to construction injury prediction. *Automation in Construction*. 2016;69:102–14.
105. Capen E. The difficulty of assessing uncertainty (includes associated papers 6422 and 6423 and 6424 and 6425). *Journal of Petroleum Technology*. 1976;28(8):843–50.
106. Rose PR. Dealing with risk and uncertainty in exploration: how can we improve? *AAPG Bulletin*. 1987;71(1):1–16.

107. Tversky A, Kahneman D. The framing of decisions and the psychology of choice. *Science*. 1981;211(4481):453–8.
108. Gustafsod PE. Gender differences in risk perception: theoretical and methodological erspectives. *Risk Analysis*. 1998;18(6):805–11.
109. Tixier AJ-P, Hallowell MR, Albert A, van Boven L, Kleiner BM. Psychological antecedents of risk-taking behavior in construction. *Journal of Construction Engineering and Management*. 2014;140(11):04014052.
110. Hallowell MR, Gambatese JA. Qualitative research: application of the Delphi method to CEM research. *Journal of Construction Engineering and Management*. 2009;136(1):99–107.
111. National Coronial Information System. *National Coronial Information System Coding Manual and User Guide (Version 4e) 2018*. Available from: www.ncis.org.au/ncis/documents/ncisdocuments.do.
112. Holloway I, Todres L. The status of method: flexibility, consistency and coherence. *Qualitative Research*. 2003;3(3):345–57.
113. Sandelowski M, Barroso J. *Handbook for Synthesizing Qualitative Research*. New York: Springer Publishing Company; 2006.
114. Vaismoradi M, Turunen H, Bondas T. Content analysis and thematic analysis: implications for conducting a qualitative descriptive study. *Nursing & Health Sciences*. 2013;15(3):398–405.
115. Braun V, Clarke V. Using thematic analysis in psychology. *Qualitative Research in Psychology*. 2006;3(2):77–101.
116. Elo S, Kyngäs H. The qualitative content analysis process. *Journal of Advanced Nursing*. 2008;62(1):107–15.
117. Hsieh H-F, Shannon SE. Three approaches to qualitative content analysis. *Qualitative Health Research*. 2005;15(9):1277–88.
118. Behm M. Relevancy of data entered into riskmaster. *NCDOT Research Project 2009–10*. East Carolina University; 2009. Available from: https://connect.ncdot.gov/projects/research/RNAProjDocs/2009-10FinalReport.pdf.
119. Moreno JL., editor. The Sociometry Reader. *Canadian Journal of Economics and Political Science/Revue canadienne de economiques et science politique*. Free Press; 1960;28(2):318–319, May 1962. Available from: https://doi.org/10.2307/139208.
120. Haythornthwaite C. Social network analysis: an approach and technique for the study of information exchange. *Library & Information Science Research*. 1996;18(4):323–42.
121. Wasserman S, Faust K. *Social Network Analysis: Methods and Applications*. Cambridge: Cambridge University Press; 1994.
122. Freeman LC. A set of measures of centrality based on betweenness. *Sociometry*. 1977:35–41.
123. Zhou Z, Irizarry J, Li Q. Using network theory to explore the complexity of subway construction accident network (SCAN) for promoting safety management. *Safety Science*. 2014;64:127–36.
124. Alsamadani R, Hallowell M, Javernick-Will AN. Measuring and modelling safety communication in small work crews in the US using social network analysis. *Construction Management and Economics*. 2013;31(6):568–79.
125. Pryke SD. Analysing construction project coalitions: exploring the application of social network analysis. *Construction Management and Economics*. 2004;22(8):787–97.
126. Pryke SD. Towards a social network theory of project governance. *Construction Management and Economics*. 2005;23(9):927–39.
127. de Brito SR, do Socorro da Silva A, da Mata EC, Vijaykumar NL, da Rocha CAJ, de Abreu Monteiro M, et al. An approach to evaluate large-scale ICT training interventions. *Information Systems Frontiers*. 2018;20(4):883–99.

5 BIM-Based Applications for Preventing Accident Causations

5.1 BIM – A Case Study

Building Information Modelling (BIM) can have a significant impact on managing accident causations by supporting:

1) Decision making early in the procurement process. The BIM mantra "build virtually before building actually" encourages project planning and issue identification and resolution early in the procurement process. This reduces stress on the team by avoiding time and cost pressures when problems are identified during construction;
2) Resolution of 3D visualisations of a building to assist in resolving spatial issues, such as conflicts on space utilisation, relationship to restricted areas, etc;
3) Simulations of the construction process (4D) to check for clashes in construction operations and to plan temporary works; and
4) Suilt-in and editable data entered for elements of the model that can contain WHS information.

This section will be referencing a case study project to illustrate some of the benefits for safety achieved through using Open BIM on building projects. The case study is from the S1 Project (Figure 5.1) undertaken in the city of Eindhoven in the Netherlands by VolkerWessels, an international company operating in Europe and North America. It has 16,000 employees across 120 operating companies within the VolkerWessels group. While VolkerWessels undertakes projects using a range of delivery approaches and contractual arrangements, the S1 project is useful as a case study as it was undertaken largely through VolkerWessels' operating companies where BIM could be used in a consistent, cohesive way from project initiation through to in-use management. While the focus is on the health and safety issues around the use of BIM, the initial discussion briefly covers how VolkerWessels uses BIM to achieve results.

VolkerWessels has been implementing BIM for 12 years. Initially, they used a fairly standard centralised technology uptake model, but after five years, the penetration of BIM was low, being utilised by about 20% of the staff. The approach was then changed to distribute the use of BIM across the operating companies so that

DOI: 10.1201/9781003224853-5

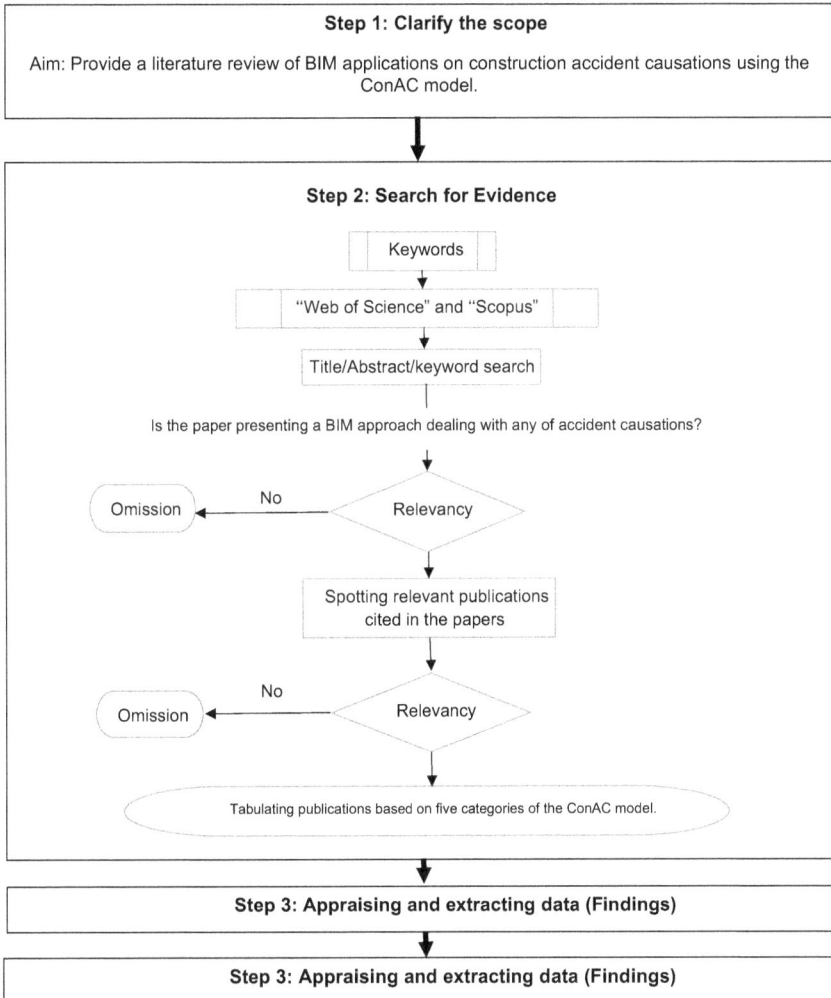

Figure 5.1 The S1 Project

Source: (1)

BIM was used by the personnel actually delivering the results, rather than checking the results from a centralised service. This led to widespread use of BIM. There is a specialist group that trains and advises the operating companies, but the operating companies are responsible for the BIM delivery. This approach has the benefit that the BIM is delivered by people who are intimately associated with the project and who have a vested interest in producing the best result. From a safety perspective, BIM models are produced by personnel who are able to apply their holistic project knowledge and experience on the ground to identify and mitigate potential safety issues on the spot in a timely manner. This ensures that the BIM is useable in the field. VolkerWessels supports the Open BIM approach, which promotes software vendor independence and the use of the most appropriate software tools as necessary.

The case study building is the S1 Building, located in Eindhoven, The Netherlands. The building is approximately 70 m tall, with an area of 11,000 m², and contains 105 apartments. The building was designed in 2018 and constructed between 2019 and 2020. Fifteen companies collaborated to produce more than 40 BIM models for the project. These were used from project initiation through design and construction and are now employed in the facility management of the building. Separate BIM models were produced for the various disciplines, which were then federated as required.

There were several aspects to the employed strategy:

1) Working from small to large scale; concentrating on resolving issues within a close context first and then scaling this up to a full model avoids being buried under complexity;
2) Using the models as the primary working method, and only generating 2D visualisations from the model when necessary;
3) Holding regular coordination meetings to resolve inter-disciplinary issues. These are held on site once construction commences so that issues can be resolved expeditiously. This also ensures that the team is up to date with on-site progress;
4) Making the models as widely available as possible across the project team, with one-click access from their devices;
5) Providing daily updates on the models.

There are several technical issues that need to be resolved to successfully use BIM to support accident prevention. Firstly, the various participants in the procurement process need to be able to share information that is relevant to other participants. Since there is no one software vendor that provides all of the necessary software tools to support BIM processes, and there are a number of options within each discipline, a system must be used that supports interoperability between software platforms. The Open BIM system was used on S1, with the software tools used by the design team shown in Figure 5.2. The central repository at the top of the figure is the Common Data Environment (CDE) where all shared information is stored.

Communication across the procurement team is important to coordinate resolution of safety issues. The BIM collaboration format (BCF) provides a mechanism for capturing issues amongst the team (Figure 5.3).

An important part of pre-construction planning is site layout – crane locations, onsite vehicle paths and material laydown areas need to be placed and relocated through the entire construction process. The appropriate crane load-carrying capacity and boom length can only be decided once the crane location has been selected and the size and weight of maximum lift components defined. Crane lifting paths also need to avoid going over workers (Figure 5.4).

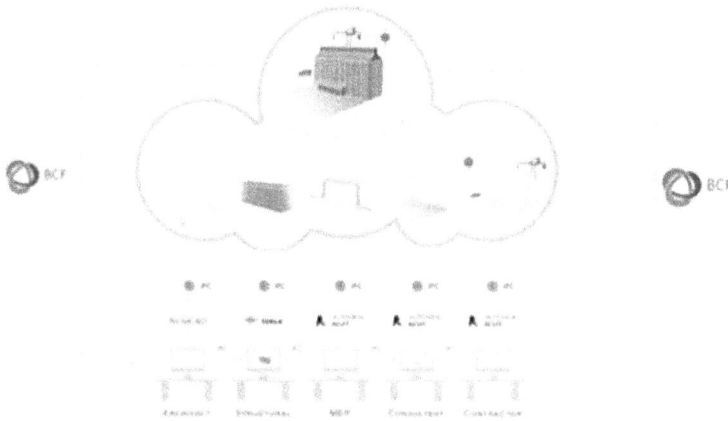

Figure 5.2 Software tools used by the design team
Source: (1)

Figure 5.3 BCF as the unifying issue tracking mechanism for S1
Source: (1)

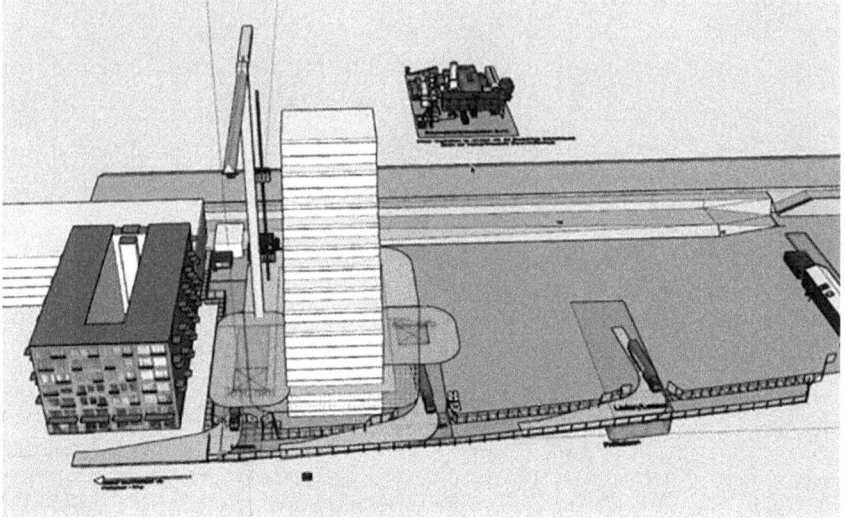

Figure 5.4 Pre-construction site layout planning
Source: (1)

As mentioned earlier, 4D modelling combines construction activity scheduling with a 3D view of the building to provide a simulation of the construction process (Figure 5.5). Overall, the main motivation is to reduce overall construction time, but from a safety perspective, 4D modelling supports checking for accident causation factors among construction activities. This can include checking for crane boom intrusion into unsafe areas, lifting paths that cross over the top of people and unsafe material storage safe locations.

A simple example from the S1 project shows that consideration of accident causation factors during the design process can reduce risk and improve the efficiency of on-site operations. The façade of the S1 project consisted of precast concrete panels which were mounted on the reinforced concrete structural frame. Scaffolding needed to be anchored at regular intervals to the building to prevent collapse. The scaffolding sub-contractor provided 2D drawings of the required locations of the scaffolding anchors, which had to penetrate the precast concrete façade panels (Figure 5.6). There were hundreds of anchors required. If any were missing, then holes would have had to be drilled through the panels at whatever height they were at. This would introduce the risk of damaging the panels, but also would require the potentially dangerous operation of drilling holes in panels located 60–70 m about ground level. The anchor points were represented diagrammatically in the BIM model. Clash detection algorithms between the anchor points and the façade panels were then used to check that there were anchor points at all required locations and that these were included in the workshop drawings for the precast concrete panels (Figure 5.7).

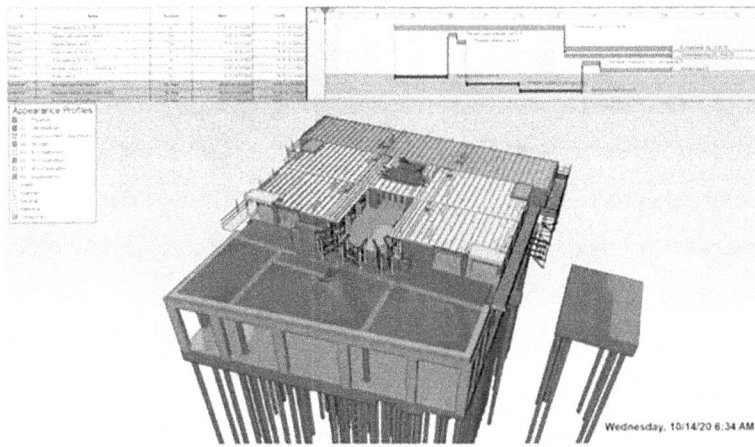

Figure 5.5 4D use of BIM – combining time and location to check for accident causations. Note the scaffolding erection process in the centre of the floor plate

Source: (1)

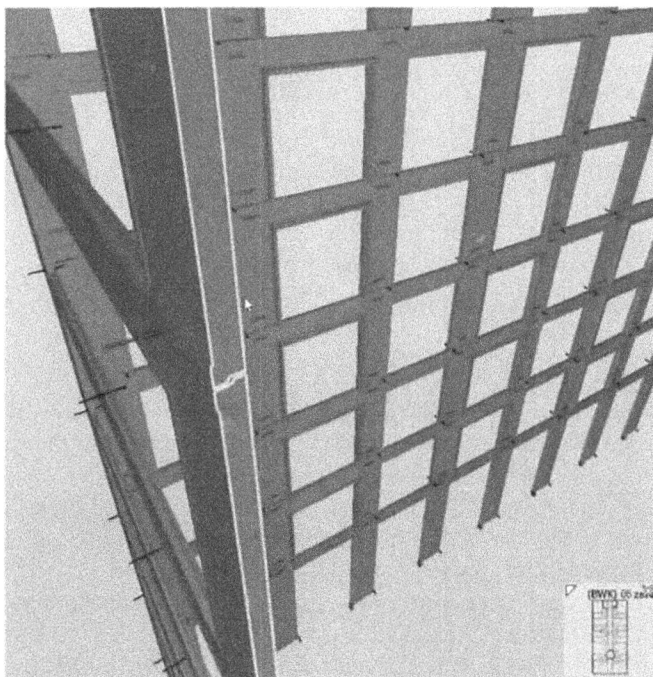

Figure 5.6 Locations of scaffolding anchor points with respect to the façade panels and structural frame

Source: (1)

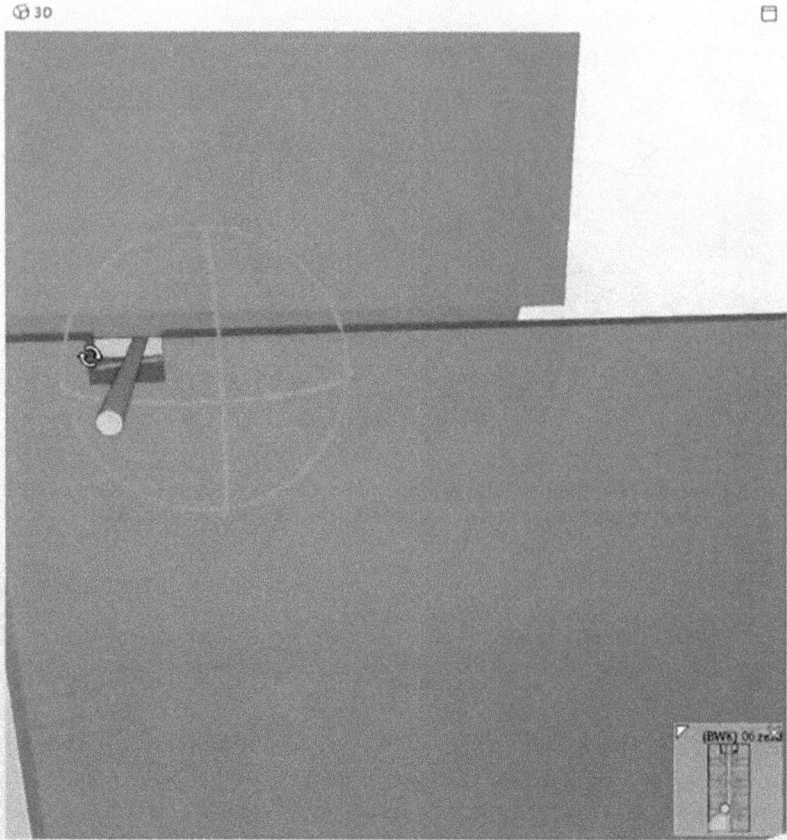

Figure 5.7 Checking required location of anchor points against precast panels
Source: (1)

On subsequent projects, VolkerWessels has built on its learnings from the S1 project to provide additional information to crane drivers and to the gatekeepers who monitor the entrances to the site and direct vehicles within the site.

A display is mounted in the crane cabin to allow the crane driver to see the location in the BIM visualisation to determine where the current load needs to be placed on the actual construction site (Figure 5.8).

The workers on site can also access the BIM visualisation to ensure correct placement of the elements (Figure 5.8).

A camera is also mounted on the crane boom to provide an improved view to the crane driver. The images are leveraged by creating point cloud images of the construction site twice a day (Figure 5.10). The major motivation for this is to provide information in the case of warrantee claims. However, an additional use would be to provide accurate information if there is a significant safety incident.

Gatekeepers play an important role in directing vehicles on site to the required location. The BIM visualisation is updated on a daily basis, allowing the gatekeeper

Figure 5.8 Crane driver using BIM to assist in placing loads

Source: (1)

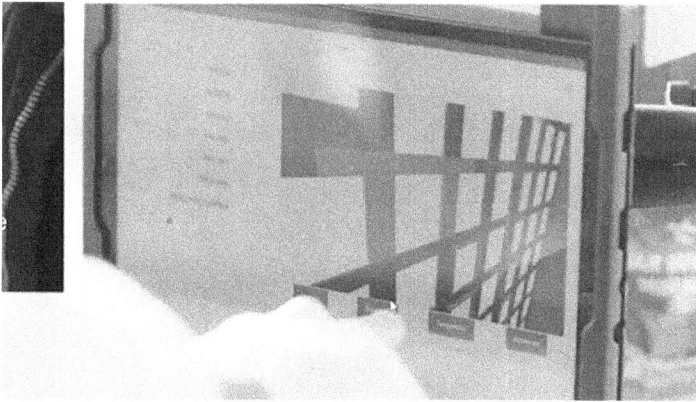

Figure 5.9 Cross-checking the actual location against the BIM guide during installation

Source: (1)

Figure 5.10 Point clouds generated twice a day can provide support for resolution of safety incidents

Source: (1)

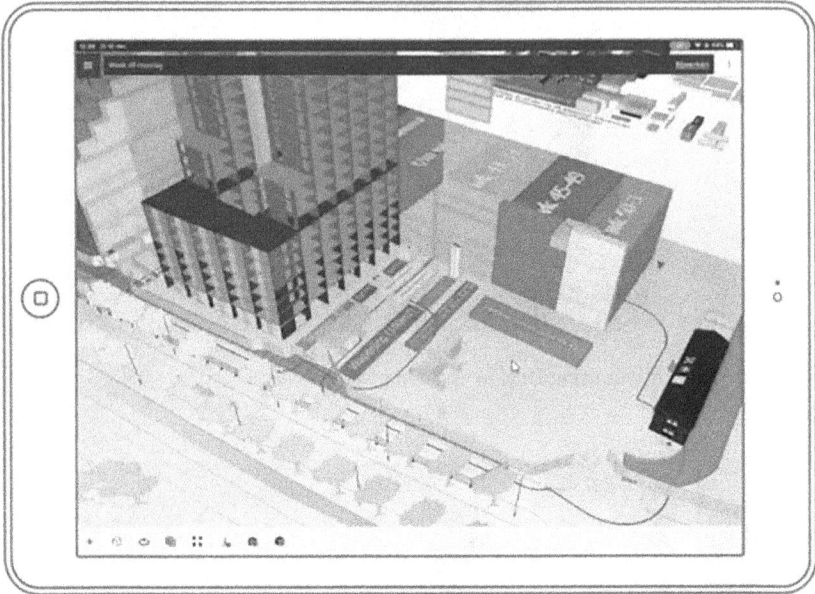

Figure 5.11 Gatekeeper's view of the BIM model

Source: (1)

to identify the most appropriate routes to delivery zones for vehicles given the changing site situation (Figure 5.11).

In most projects, neither crane drivers nor gatekeepers have access to these levels of information. Both parties report higher levels of job satisfaction through improved levels of information access and delegation of some levels of decision making. This may improve potential mental health issues by allowing on-site operatives to use their full potential.

5.2   BIM and Accident Prevention

After describing the WHS benefits of employing BIM in construction projects, this chapter focuses on investigating the potential applications of BIM towards WHS. A literature review by Guo, Yu (2) indicated that since the year 2012, the number of scientific publications on this topic had increased considerably (more than doubled). This study showed that BIM-based approaches support training, identify job hazard areas and monitor construction sites (2). In another recent holistic review paper, Zou, Kiviniemi (3) examined BIM-based approaches through the lens of the risk management process.

In the current study, a holistic review of the existing literature was conducted on published journal papers from 2007 to 2022 and on the references given within the papers. Unlike Zou, Kiviniemi (3), who considered BIM technology, and Guo, Yu (2), who examined literature for visualisation aspects of BIM, the review for the current study was focused on relevant knowledge regarding BIM's implications on WHS practices and related them to a wide range of accident risk drivers presented in the ConAC model. The aim was to facilitate the identification of the effectiveness of each application on the identified types of risk drivers.

As illustrated in Figure 5.12, this study was based on a four-step process for a systematic literature review, following the approaches taken by Pawson, Greenhalgh (4) and Chong, Lee (5).

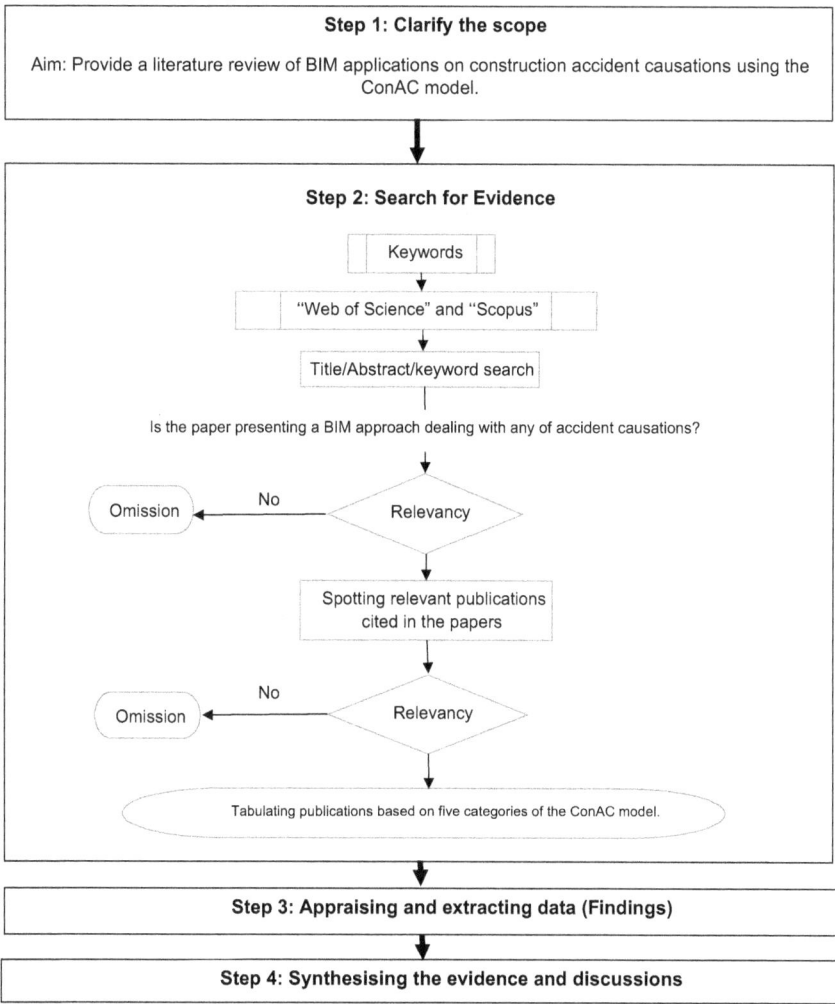

Step 1: Clarify the scope

Aim: Provide a literature review of BIM applications on construction accident causations using the ConAC model.

Step 2: Search for Evidence

Keywords

"Web of Science" and "Scopus"

Title/Abstract/keyword search

Is the paper presenting a BIM approach dealing with any of accident causations?

Omission ← No — Relevancy

Spotting relevant publications cited in the papers

Omission ← No — Relevancy

Tabulating publications based on five categories of the ConAC model.

Step 3: Appraising and extracting data (Findings)

Step 4: Synthesising the evidence and discussions

Figure 5.12 Literature review processes to categorise the literature findings

As a result, 81 articles were found to be relevant to BIM applications related to accident causations. To ensure that all related articles are involved in the data set, Webster and Watson (6) recommend taking into account the reference lists of the collected articles, as illustrated in Figure 5.12. By taking this approach, eight additional articles were identified to be relevant, bringing the overall number of articles to 89. After finalising the search for evidence, the articles related to BIM applications were subjected to content analysis to identify their feasibility for construction WHS (Table 5.1).

As a result of the literature review, these studies were found to be mainly divided into six topics: 3D tools for preliminary risk assessment, automatic/semi-automatic model checkers, 4D (3D+time) construction planning, knowledge management system, AR and VR and RTLS. Table 5.1 describes the findings of the previous studies related to each topic and concludes with BIM's WHS effectiveness for each type of accident risk driver.

5.3   3D Visualisation Tool for Preliminary Risk Assessment

Data-rich 3D models are the least technical tools for extracting safety-related information by monitoring the facility for better and faster imagination of the working environment and modifying potential unsafe conditions (7).

3D BIM models are more helpful than paper-based 2D models during risk assessment sessions held before construction between the designer, contractor and other project stockholders. The main purpose of these sessions is to assess safe constructability and potential risk within the proposed project. This takes place during different design phases, providing possibilities for PtD, where the designer can use the construction knowledge of the contractor to produce an inherently safer model (8, 9). 2D drawings do not effectively support the visualisation of a potential hazard, and therefore do not promote communication between team members regarding technical safety modifications and provisions. Hadikusumo and Rowlinson (10) described the benefits of 3D visualisation for WHS management as "what you see is what you get".

Story and Mantle (11) conducted a trial study to determine how a risk matrix could be combined within a BIM package, namely Autodesk Revit Architecture. Although the study focused on the operation phase, the applied method could be utilised in other phases of a project's life cycle to calculate and visualise the risk values of potential hazards. This type of manual embedding of risk information within BIM models, although time-consuming, can allow for the visualisation of hazardous areas by all workers, from top experts to those with inadequate hazard recognition knowledge (12).

Moreover, there is a higher risk of injury for craftsmen new to the workplace until they understand the environment of the work site (12). Using BIM, they can learn in a faster and a more constructive manner about the layout and complexities of the site (12). For example, a BIM model can show storage areas for hazardous materials, existing structures, traffic control arrangements, overhead and underground utilities, crane swing radius, etc.

Table 5.1 BIM-enabled options that can positively influence accident risk factors

		3D visualisation tool for preliminary risk assessment	Automatic/semi-automatic model checkers	4D construction planning	knowledge management system	AR/VR	RTLS
Work team	Communication	*		*	*	*	*
	Capabilities (knowledge/skill)				*	*	*
	Actions/behaviours					*	*
	Health/fatigue				*	*	*
	Supervision	*			*	*	
	Attitude/motivation				*	*	
Workplace	Local hazards (e.g., harsh climate)	*			*	*	*
	Site conditions (e.g., site topography)				*		*
	Working environment (e.g., dust)		*				
	Site constraints (e.g., access and egress)	*		*		*	*
	Work scheduling	*		*			*
	Housekeeping	*	*	*			
Material/ equip	Suitability of material/equipment	*	*	*			*
	Condition and usability of material/equipment						*
	Design specs of material/equip						
	Availability of material/equipment						
Originating influences	Permanent work design	*		*	*	*	
	Project management	*		*	*	*	*
	Construction processes	*	*	*	*	*	
	Safety culture				*	*	
	Risk management	*	*	*	*	*	

5.4   BIM and Prevention Through Design & Planning

Existing safety guidelines, rules and best practices can be examined using 3D BIM-based design and planning models via an automated safety rule checking system. The advantage of automatic identification of dynamic hazardous conditions is the ability to detect their location in a virtual model, and then automatically or semi-automatically advise preventive measures (13). Eastman, Lee (14) defined the term "automatic rule checking" as employing a software platform for assessing a design according to configurations of its objects by encoding rules so that the software automatically produces the results of the design, such as pass, fail, warning, or unknown.

Hammad, Setayeshgar (15) defined the rules for a model checker program based on the Safety Code of Quebec Provence in Canada to identify the open edges of buildings. In addition, this model automatically places the temporary fencing within a work breakdown structure at the planning stage for temporary works and confined spaces. Sulankivi, Zhang (16) developed a BIM-based automatic safety rule–checking prototype in order to identify the safety risks that were not recognised by designers. This prototype simulated construction activities and sequences and identified hazardous locations and situations. Zhang, Teizer (17) translated the properties of fall protection, extracted from OSHA guidelines and other best practices, into rule-based algorithms and incorporated them into a rule-checking tool to detect the best location for fall protection. The practicality of the tool was examined via Tekla Structures software.

Oh, Gross (18) defined three major features for a BIM-supported PtD software based on previous studies on the design process. The first feature is concerned with conditions related to activating the intervention mechanism. Systems incorporate an active critiquing mechanism with constant monitoring of the design process and creating feedback, while a passive mechanism provides feedback upon request. In recent studies, the timing of PtD feedback has only been considered at the end of the design phase (17, 19).

The second feature deals with the nature of feedback. The applications for rule-checking in current practice identify the safety errors in BIM-based models and create comments on them. The tools can also suggest opportunities to improve the health and safety of workers, even when the design is errorless.

The third feature concerns the form of a report that reflects the created feedback. Two tools, text messages and graphical markers, have been used to highlight the location of the errors and possible improvements in recent studies (17, 19). Designers can better understand safety hazards when the nature and risk of the hazards are further explained by text messages. Components of the structure that have the possibility of hazard are marked in the 3D model and designers can then work on correcting them.

A comprehensive study by Solihin and Eastman (20) classified the rules used for automated model-checking into four levels according to the complexity of the required data. A safety input can only be checked if it applies to a set of data and components available in the BIM model.

In a recent study, Vakilinezhad, Dias (21) pointed out eight main barriers they faced when converting OSHA regulation Part 1926 into rules for model checkers, including:

- the vagueness of the narratives in the regulation;
- the outsized content of some narratives in the regulation;
- the non-consistency of mapping terminology in the regulation;
- the existence of narratives that could not be tracked in a BIM model (inspection process, human behaviour, etc.);
- the absence of required components within the BIM models;
- the absence of required attributes within the BIM components;
- the inflexibility of the model checker to define new rulesets; and
- the missing component decompositions (e.g., subparts of the manually designed components, such as scaffolds and ladders [defined as a discrete accessory in IFC format] are not recognised by model checkers).

5.5   4D BIM and Construction WHS

Integrating a safety plan with the project schedule to improve the WHS performance of an organisation is not a new approach (22, 23). Mallasi (24) employed 4D CAD to identify construction site congestion points and possible safety hazards. This was one of the earliest attempts to create dynamic construction schedules and visualise the construction environment to facilitate the decision-making process of planners. The developed tool runs a critical space-time analysis on the 4D model to quantify the congestion level in a workplace.

As one of the world leaders in the construction industry, the Skanska construction group used 4D BIM in the construction of the Barts and London Hospitals to investigate the possibilities of trip hazards during the installation of new services (25). The VTT Technical Research Centre of Finland employed the BIM software Tekla Structures as a central technology for a systematic approach towards safety communication and management (25). The presented approach aimed to explore how methods for 4D construction site layout and safety-related scheduling could be integrated using BIM. Sulankivi et al. (2010) carried out a test on the Mäntylinna residential building project from the Skanska group and linked the construction schedule with a temporary structure, building components and site equipment. However, due to several missing temporary structures, specific building components, site equipment and machinery within the BIM software library, the contractor faced difficulties when designing them. Sulankivi, Kähkönen (25) indicated that 4D BIM can improve WHS by closely integrating safety concerns with the construction planning, visualising safety plans and site layout in more detail, effectively updating site status information and facilitating safety communication by alerting site personnel about future risks and devised control measures.

Irizarry and Karan (26) integrated the use of BIM and geographic information systems (GIS) and proposed a GIS-BIM model to assist with the identification and optimisation of the ideal locations of tower cranes. In this work, BIM software was first used to generate geometric information of the construction site, and the GIS model then extracted data from the BIM to determine the proper combination of tower cranes for location optimisation (Irizarry & Karan, 2012). The analysis output linked to the BIM platform can suggest one or more possible areas, including all supply points and demand.

The primary design of a structure may encounter several variations during construction due to constructability concerns. Thus, the dynamic nature of construction projects requires a timely analysis of the structural safety to foresee and prevent structural failures. Hu and Zhang (27) developed the 4D-GCPSU 2009 tool, which can run time-dependent structural analysis. Such analysis is facilitated by the structural information already embedded in the 4D BIM models, including loading conditions, materials, local axes, profiles (such as inertia moment, centroid, area, etc.), structural element types, etc. During construction of the Wuhan Mingdu Metro Station project, Zhou, Ding (28) used 4D BIM to conduct a number of safety risk analyses, including assessments of landslide hazards in excavation procedures, crane operation failures, retaining pile collapses, water in-rushing, seepage caused by fractured retaining piles, bracing structure deformations in the excavation procedures, bottom heaves caused by artesian water and damage to adjacent buildings.

Zhou, Ding (28) highlighted the major benefits of 4D BIM for safety controls as:

- identification of unsafe construction procedures by simulating the process;
- visualisation of evolving patterns of safety risks that vary both spatially and temporary;
- safety status monitoring of each component of updated 4D models; and
- identification and marking of hazardous, providing suitable control measures.

Wang, Zhang (29) used 4D BIM to develop an approach for real-time online safety analysis of a time-dependent underground cavern construction project. The model employed C# and the OpenGL (Open Graphics Library) Tao Framework to integrate construction progress information, numerical simulation data (obtained from Abaqus software), monitoring information and geological information.

Measures to control environmental factors, such as air quality and dust level, that influence workers' health can be integrated with the 4D BIM model. Because BIM models enable indoor air quality analysis, Altaf, Hashisho (30) developed a method to analyse air pollutant concentration during construction activities. The proposed model was tested by measuring the level of air pollutants during drywall sanding activity and modifying the construction schedule by embedding control measures.

5.6   BIM-Based Knowledge Management System and Construction WHS

An important feature of BIM models is that they represent a repository of various types of information that prevent the storage of bulky, paper-based documentation. These documents may include a large number of safety management requirements, such as the health condition of the workers, safety of equipment and machinery, records of accidents, etc. Goedert and Meadati (31) extended the practice of BIM to construct a sole repository of facility data to be used after the construction phase. Fruchter, Schrotenboer (32) introduced BIM as a rich multimedia building knowledge model and argued that everything about the building could be communicated within a building knowledge model.

Ganah and John (33) proposed integrating tool-box talk for facilitating safety communication among different construction stakeholders and BIM. Within the proposed approach, WHS concerns appear as a separate property attached to BIM components and activities, and every alteration, change, and modification is stored with a time track to represent real safety concerns at each specific point in time. In a more meticulous approach, Zhang, Boukamp (34) developed a construction safety ontology to systematically manage safety knowledge in construction projects. In this system, job hazard analysis forms that are traditionally used to facilitate pre-task safety communication are embedded within the BIM model. This includes three main ontology models of construction: Safety, construction process, and construction product models. In a similar study, Ding, Zhong (35) developed an ontology-based platform in a BIM environment and using web technology for effective management of safety knowledge.

Kim, Lee (36) developed a system that stores past accident cases in a relevant BIM component or activity and retrieves them based on the user's search, where a similar activity is performed and a possible safety risk can be identified from past experiences. The query system in this model retrieves past accident information according to work, work conditions and labourers. In a more advanced approach, Shen and Marks (37) simulated near-miss information using BIM to be stored and used for safety risk assessment. In this study, they simulated reported near-misses by workers in a BIM platform, storing the data and facilitating further analysis by safety managers to identify the frequency and severity of the reports.

Zou, Kiviniemi (38) employed a data mining method to develop a knowledge-based safety risk database in the construction of a bridge project. Zou, Kiviniemi (38) argued that the current safety risk knowledge transition process within the industry is fragmented and insufficient, and lessons to be learned face several barriers that can be transferred from one project to another similar one.

5.7 BIM-Based AR/VR and Construction WHS

The term "second life environment" refers to a virtually interactive setting where users can freely explore and take action within various simulated scenarios where such circumstances in a real world would be costly and involve serious safety hazards. In the construction industry, such an environment can be enabled through VR and AR technologies where BIM facilitates creating the settings (39).

According to Steuer (40), VR is a technological system comprised of a virtual real-time scenario creation tool, a body motion tracker tool, and a head-mounted stereoscopic display.

Reiners, Stricker (41) described AR as the technology of integrating pictures captured from the real world with virtual objects. These pictures can be captured using a camera, head-mounted scanners or even a direct view of a person from the real world.

One of the earliest attempts to employ VR for improving WHS communication was employed during pre-construction risk assessment sessions in a collaborative atmosphere where different project stockholders investigated PtD concerns (10). The developed model comprised three main aspects, including a PtD database, a data-enriched building model, and VR functions. Within the created virtual project, users could walk in the environment of the model from a first-person view, investigate potential safety risks and address suitable control measures. Sacks, Whyte (7) indicated that designers navigating in the virtual environment of the designed structure together with a safety professional provides the setting for a collaborative conversation to identify possible hazards and discuss feasible PtD solutions.

Using a second life environment can improve safety-related training under conditions where mistakes by the trainee do not result in any type of injury, and actions can be analysed in more detail through the sensors attached to the trainee. In this form, gaming technologies support creating interactive and purposeful scenarios. H. Li, Chan (42) described gaming technology as an engine comprising several technical aspects such as simulation, digitalisation, instructiveness, and intelligence and provides multi-user and network-based platforms. H. Li, Chan (42) customised Wii controllers and keyboards to be used for training crane tower operators. Sacks, Perlman (39) compared WHS knowledge learned from traditional classroom training with that obtained using a 3D immersive VR power-wall training method and concluded that VR-based training was more efficient for the identification and prevention of safety risks.

In an innovative approach, Albert, Hallowell (43) developed a dynamic and interactive augmented virtual environment using BIM models to develop workers' hazard recognition skills through risk-free learning. They simulated two diverse projects that involved hazards related to the maintenance of oil and gas plants and the construction process of a fluff pulp manufacturing plant. The results of the study demonstrated significant improvement in workers' hazard recognition skills. This was the first attempt to develop a comprehensive training protocol to improve hazard recognition skills and the first effort to measure the impact of a human-centric augmented virtuality tool on adult learning.

Bosché, Abdel-Wahab (44) integrated state-of-the-art virtual reality goggles with an innovative six-degrees-of-freedom head pose tracking system to facilitate walking through training in the VR model. The preliminary results of the study revealed the efficiency of the method in portraying the realistic environment of the worksite, especially for working at height. To providing a more realistic environment in VR models for construction training, Lu and Davis (45) integrated audio features that exist in various scenarios.

Zhao, McCoy (46) developed a mobile virtual reality prototype to establish a safety culture for electrical workers in their daily practices. The results of the study validated the application of the prototype for improving the safety culture among workers.

5.8 BIM-Based RTLS and Construction WHS

In recent years, several studies have examined the practicality of the RTLS concept, which integrates navigation systems (e.g. global positioning systems, RFID, Wi-Fi, etc.), sensors, and BIM into a hub to prevent accidents caused by unsafe proximity of pedestrian workers resulting in crashes with construction equipment, machinery and crane towers (47–51). RTLS contributes to the prevention of several accident causations by improving work team performance (communication, supervision and project management), workplace conditions (housekeeping and construction process) and equipment/machinery interactions with on-foot workers. Although the accuracy of RTLS in its current state and the financial justification for employing it in small and medium-sized projects might be questionable, it can provide a higher level of digitalised and automated safety controls.

Lee, Lee (52) criticised the non-applicability of GIS to the indoor areas of construction projects and developed an RTLS method based on RFID technology that indicates the location of an object with an average error of 86.50 cm. The developed method has three main components: Assistant tag to maintain the availability of a received signal, chirp spread spectrum technology for wireless networking, and a localisation method for the time of arrival. Similarly, N. Li, Li (53) used RFID within an RTLS algorithm that could determine the location of stationary and mobile objects with a mean accuracy of 1.94 ± 0.17 m and 1.42 ± 0.49 m, respectively.

Cheng and Teizer (54) developed a framework to employ RTLS for the real-time monitoring of ongoing activities on a construction site. The result of testing the framework showed that important WHS information could automatically be captured in a live form and safety managers could remotely inform workers and equipment operators about the situation.

Naticchia, Vaccarini (55) introduced an infrastructure-less ZigBee-based model for use by site inspectors to track the location of site workers. Soleimanifar, Shen (56) examined the applicability of received signal strength methods in real-time case studies and concluded that a profiling-based, received signal strength method combined with noise filtering algorithms resulted in a high level of accuracy, identifying a point within 1–2 m of the actual position of a target. Recently, H. Li, Chan (48) adopted a chirp spread spectrum technique to capture the position of objects and concluded that this method provided higher accuracy and lower cost of implementation than RFID and ultra-wideband (UWB) technologies. A study by Cheng, Migliaccio (57) examined the integration of RTLS with sensors that indicated the fatigue level of the workforce by measuring their physiological statuses.

Determining the location of workers and equipment facilitates the development of models that define an approximate zone for potential WHS risks. Ren and Wu (58) developed an anti-collision system based on RTLS to prevent crashes between mobile cranes and solid objects on construction sites. Teizer and Cheng (59), Wang and Razavi (60), and Isaac and Edrei (61) retrieved the movement path of construction equipment and on-foot workers using RTLS technologies to develop approximate safety zones and prevent collision hazards.

5.9   Critical Applications of BIM Against the ConAC Model

To confirm the findings of the literature review and identify approaches of BIM with higher efficiency than that of the ConAC model, a semi-structured interview method was adopted. According to Robson (62), in semi-structured interviews, potential and relevant sub-questions can be prepared to facilitate a meaningful conversation and ensure that all aspects of the topic are covered. The semi-structured interview was deemed the most suitable approach for this exploratory study because there are limited prior studies regarding the adoption process of BIM for construction WHS. As stated by Seidman (63), a semi-structured interview allows researchers to gain and discover new subjective information.

In selecting the interviewees, it was imperative to ensure that the participants were able to provide industry-specific and realistic information based on their experience in the industry. As a result, the goal was to find a variety of experts who were both consultants and contractors and who employed BIM on a daily basis. The experts included a project manager, a BIM manager, a BIM coordinator, the national HSE manager, and architects. For ethical reasons, the interviewees are presented anonymously. In addition to providing a select sample representative of a variety of expertise, the sample size met the criteria that no new codes emerged in the last two interviews.

In the interviews, the interviewees were asked to express their views about the applicability of each BIM application to control the critical accident causations of workers' actions and behaviours (WT1), worker capabilities (WT2), immediate supervision (WT4), worker health and fatigue (WT5), equipment condition (E3), permanent (and temporary) work design (OI1), construction process (OI3) and risk management (OI5). They were also asked to mention any other BIM-based approaches that could be applicable for controlling critical causations.

The interviewees believed that the use of technology could support an improvement in workers' perceptions of the risks and outcomes of their actions. Virtual training and education using VR, AR, and/or BIM models were addressed as a means to enhance a worker's ability to identify the potential risks on construction sites. They believed that this approach was effective in the training of workers by improving the safety culture of the workers, developing safer work method statements and communicating safety issues to those workers with language barriers. However, Interviewee 2 mentioned that,

> Workers' actions and behaviours are totally a cultural issue and technology has the least effect in [*sic*] improving it. Workers decide to not follow a safe work method in their daily practice and performing [*sic*] a job in an unsafe manner to spend less time and energy.

The interviewees agreed that prevention at the design stage using automated or semi-automated model checkers could allow design teams to identify unsafe designs based on predefined safety rules. They also pointed out that BIM models provided a clearer platform for safety in design meetings involving the client, consultant and contractors.

Interviewee 1 mentioned that "BIM facilitates identifying and clash detecting of the static equipment, like scaffoldings or guardrails in the sites. This method, however, requires extensive work on modelling of the temporary equipment".

The participants believed that 4D modelling and risk assessment at the planning stage constituted an effective approach to provide a simulation of the site activities and modify the work method statements. Interviewee 4 pointed out that:

> While 4D BIM is capable of visualising the activities, it is also capable of providing risk analysis, that at the current stage, large companies have produced their own version of risk analysis of the dynamic equipment, like machinery, and assess their suitability for the tasks.

Participants mentioned that, by nature, 4D BIM was a tool used to develop the construction schedule, and in this manner, it facilitates recognising the required equipment at the right time. Participants also repeatedly mentioned that 4D BIM could provide a platform for assessing sub-contractors' level of understanding in safely performing their contracts.

Monitoring functions of the BIM applications were mentioned as the most effective tools, among others. Although a range of approaches were mentioned, these primarily referred to the digital monitoring of the sites. Participants stated that although using sensors to pinpoint the location of the workers and alerting them if they were entering an unsafe proximity of a hazard would not 100% prevent workers from putting themselves at risk, it could prevent a large proportion of the accidents.

Interviewee 5 stated that:

> BIM 360 Field as a practical application for systematic monitoring of the site by the site supervisors. Similar applications provide a collaborative platform for the project team to communicate the existing hazards related to site condition, condition of the equipment, reporting sub-contractors safety performance, and it improves [the] safety culture among the members of the organisation.

Most of the interviewees agreed that sensors attached to the body of the workers and connected to a BIM model could measure the health condition of the workers and report this to the management team, as well as alerting the workers. In addition, interviewee 3 outlined that:

> Environmental sensors can measure the severity levels of the weather and provide immediate alerts in indoor and outdoor areas. Connecting these sensors to a model can pinpoint the location of hazards, and consequently help the management team to come up with proper action.

Interviewee 5, an expert in laser scanning, mentioned laser scanners as another type of sensing tool that could capture real situations on-site and suggested that importing these captures to a BIM model could help to identify the variations, unsafe conditions on site, and suitability of the operating equipment.

BIM was described by the interviewees as a decent digital repository of project lifecycle safety information that can store all necessary WHS information about the project and be handed to different trade workers and sub-contractors during the maintenance of an asset.

Interviewee 4 outlined that:

> This approach is effective in transferring the special requirements related to site conditions that may not be recognised by the maintenance workers. This approach is very effective in transferring the knowledge learned to [sic] the projects to the next projects. It can identify the workers with a poor WHS performance record, equipment operating conditions, modified work method statements, and workers' health conditions.

Interviewees believed that BIM applications among material and equipment suppliers were very limited in the industry, while the use of the BIM models could assist suppliers in assessing the suitability of their supplies. 4D BIM was mentioned as a tool to leverage the on-time provision of equipment and material that could prevent accidents that take place due to their unavailability.

Last but not least, most interviewees believed that several factors contributed to accidents and that controlling these factors would result in preventing accidents. Therefore, BIM is not the ultimate solution to achieve zero accidents, but it can be a major component of the plan.

To summarise the findings from the interview sessions, 17 potential approaches were identified as means of intervention by BIM to control the causations found in the previous section and were coded to be used in the questionnaire survey (Table 5.2).

Table 5.2 Applications of BIM to control the critical accident causations, identified via interview

Approach	Code	Description
Virtual training, education and method statement development	BIMtype1_1	Training workers in an interactive virtual environment that can involve all existing risks rather than traditional classroom-based training.
	BIMtype1_2	Testing and improving workers' safety culture in situations such as reporting unsafe conditions on the site, recognising the required PPE, communicating safety concern with their colleagues, etc.
	BIMtype1_3	Developing safe work method statements for the companies and trade unions.

(Continued)

Table 5.2 (Continued)

Approach	Code	Description
Prevention through permanent and temporary work design	BIMtype2_1	Performing automated or semi-automated checking of the building designs in BIM models to avoid unsafe designs or to identify possible hazards as early as possible.
	BIMtype2_2	Performing automated or semi-automated designing and checking the designs of the on-site temporary structures using validated algorithms to avoid design related failures.
4D risk assessment at the planning stage	BIMtype3_1	Assessing the suitability of the dynamic equipment/machinery using 4D BIM simulations.
	BIMtype3_2	Planning ahead for the required materials and equipment for safer execution of the projects using 4D BIM planning.
	BIMtype3_3	Assessing workers' and sub-contractors' capabilities by conducting a walkthrough of the 4D simulations and discussing their approaches for safely performing their tasks.
Project monitoring and management at the construction stage using visualised sensing technologies	BIMtype4_1	Monitoring workers and equipment through attached location tracking sensors that are connected to a BIM model and controlled by the project management team.
	BIMtype4_2	Performing frequent monitoring of the site and equipment conditions through capturing site conditions using flying drones.
	BIMtype4_3	Communicating safety issues through software such as BIM 360 Field, which circulates information through tablets and smartphones and informs the people that are responsible to take action.
	BIMtype4_4	Monitoring workers' health and fatigue condition by using remote health monitoring sensors that can capture the location and health condition of the workers and present them in a BIM model.
	BIMtype4_5	Monitoring environmental conditions such as temperature, humidity and wind speed in indoor and outdoor locations through sensors linked to a BIM model.

(*Continued*)

Table 5.2 (Continued)

Approach	Code	Description
Digital knowledge management of project WHS information	BIMtype5_1	Using BIM as a digital library of the site conditions. An external database that stores site safety information is created and the required safety information is linked to the objects in the BIM model. This provides project/safety managers with a comprehensive understanding of the situations.
	BIMtype5_2	Using BIM as a digital library of equipment/machinery maintenance information. An external database is created to store operation and maintenance information about the equipment/machinery and linked to the relevant equipment/machinery used in the 4D BIM models.
	BIMtype5_3	Using BIM as a digital library of the workers' safety track records. An external database of the workers' health and performance records is created and linked to the 4D BIM models. This shows workers' health and safety performance for different activities.
	BIMtype5_4	Using BIM models for suppliers to assess the suitability of the materials and equipment they provide to the construction sites.

5.10   Summary

The purpose of this chapter was to demonstrate the effectiveness of BIM in the prevention of accident causes reflected in the ConAC model. A case study was presented showing the use of BIM for WHS management. Then, relevant publications and key aspects of BIM contributing to construction health and safety were identified through a systematic literature review. We identified six categories, describing their applications and limitations. Next, semi-structured interviews were conducted with a panel of experts to identify potential approaches to preventing accidents with BIM. It was concluded that BIM can contribute to construction WHS through virtual training, education and method statement development; prevention through permanent and temporary work design; 4D risk assessment at the planning stage; project monitoring and management at the construction stage using visualised sensing technologies; and digital knowledge management of project WHS information.

References

1. VolkerWessels. S1 project. *The BuildingSMART International Standards Summit and National Conference; 18–21 October 2022*. Montreal, Canada: BuildingSMART; 2022.

2. Guo H, Yu Y, Skitmore M. Visualization technology-based construction safety management: a review. *Automation in Construction.* 2017;73:135–44.

3. Zou Y, Kiviniemi A, Jones SW. A review of risk management through BIM and BIM-related technologies. *Safety Science.* 2017;97:88–98.

4. Pawson R, Greenhalgh T, Harvey G, Walshe K. Realist review – a new method of systematic review designed for complex policy interventions. *Journal of Health Services Research & Policy.* 2005;10(1_suppl):21–34.

5. Chong H-Y, Lee C-Y, Wang X. A mixed review of the adoption of building information modelling (BIM) for sustainability. *Journal of Cleaner Production.* 2017;142:4114–26.

6. Webster J, Watson RT. Analyzing the past to prepare for the future: writing a literature review. *MIS Quarterly.* 2002;26(2):xiii–xxiii.

7. Sacks R, Whyte J, Swissa D, Raviv G, Zhou W, Shapira A. Safety by design: dialogues between designers and builders using virtual reality. *Construction Management and Economics.* 2015;33(1):55–72.

8. Gambatese JA, Behm M, Hinze JW. Viability of designing for construction worker safety. *Journal of Construction Engineering and Management.* 2005;131(9):1029–36.

9. Ku K, Mills T, editors. Research needs for building information modeling for construction safety. *International Proceedings of Associated Schools of Construction 45nd Annual Conference*, Boston, MA; 2010.

10. Hadikusumo BHW, Rowlinson S. Integration of virtually real construction model and design-for-safety-process database. *Automation in Construction.* 2002;11(5):501–9.

11. Story D, Mantle E, editors. The use of BIM to enhance the management of health and safety risk. *The Construction, Building and Real Estate Research Conference of the Royal Institution of Chartered Surveyors, The Australasian Universities' Building Educators Association Conference*, Sydney, Australia; 2015.

12. Rajendran S, Clarke B. Building information modeling: safety benefits & opportunities. *Professional Safety.* 2011;56(10):44–51.

13. Zhang S, Sulankivi K, Kiviniemi M, Romo I, Eastman CM, Teizer J. BIM-based fall hazard identification and prevention in construction safety planning. *Safety Science.* 2015;72:31–45.

14. Eastman C, Lee J-M, Jeong Y-S, Lee J-K. Automatic rule-based checking of building designs. *Automation in Construction.* 2009;18(8):1011–33.

15. Hammad A, Setayeshgar S, Zhang C, Asen Y, editors. Automatic generation of dynamic virtual fences as part of BIM-based prevention program for construction safety. *Proceedings of the 2012 Winter Simulation Conference (WSC).* IEEE; 9–12 December 2012.

16. Sulankivi K, Zhang S, Teizer J, Eastman CM, Kiviniemi M, Romo I, et al., editors. Utilization of BIM-based automated safety checking in construction planning. *Proceedings of the 19th International CIB World Building Congress*, Brisbane Australia; 2013.

17. Zhang SJ, Teizer J, Lee JK, Eastman CM, Venugopal M. Building information modeling (BIM) and safety: automatic safety checking of construction models and schedules. *Automation in Construction.* 2013;29:183–95.

18. Oh Y, Gross M, Do E. Computer-aided critiquing systems. *Proc CAADRIA (Computer Aided Architectural Design Research in Asia).* Chaiang Mai, Thailand; 2008: 161–7.

19. Qi J, Issa RR, Olbina S, Hinze J. Use of building information modeling in design to prevent construction worker falls. *Journal of Computing in Civil Engineering.* 2013;28(5):A4014008.

20. Solihin W, Eastman C. Classification of rules for automated BIM rule checking development. *Automation in Construction.* 2015;53:69–82.

21. Vakilinezhad M, Dias P, Ergan S, editors. Achieving model-based safety at construction sites: BIM and safety requirements representation. *Proc of the 33rd CIB W78 Conference.* Brisbane, Australia; 2016.

22. Kartam NA. Integrating safety and health performance into construction CPM. *Journal of Construction Engineering and Management.* 1997;123(2):121–6.

23. Wang W-C, Liu J-J, Chou S-C. Simulation-based safety evaluation model integrated with network schedule. *Automation in Construction.* 2006;15(3):341–54.

24. Mallasi Z. Dynamic quantification and analysis of the construction workspace congestion utilising 4D visualisation. *Automation in Construction.* 2006;15(5):640–55.

25. Sulankivi K, Kähkönen K, Mäkelä T, Kiviniemi M, editors. 4D-BIM for construction safety planning. *Proceedings of W099-Special Track 18th CIB World Building Congress.* Salford, UK; 2010.

26. Irizarry J, Karan EP. Optimizing location of tower cranes on construction sites through GIS and BIM integration. *Journal of Information Technology in Construction (ITcon).* 2012;17(23):351–66.

27. Hu Z, Zhang J. BIM- and 4D-based integrated solution of analysis and management for conflicts and structural safety problems during construction: 2. Development and site trials. *Automation in Construction.* 2011;20(2):155–66.

28. Zhou Y, Ding LY, Chen LJ. Application of 4D visualization technology for safety management in metro construction. *Automation in Construction.* 2013;34:25–36.

29. Wang C, Zhang S, Du C, Pan F, Xue L. A real-time online structure-safety analysis approach consistent with dynamic construction schedule of underground caverns. *Journal of Construction Engineering and Management.* 2016;142(9):04016042.

30. Altaf MS, Hashisho Z, Al-Hussein M. A method for integrating occupational indoor air quality with building information modeling for scheduling construction activities. *Canadian Journal of Civil Engineering.* 2014;41(3):245–51.

31. Goedert JD, Meadati P. Integrating construction process documentation into building information modeling. *Journal of Construction Engineering and Management.* 2008;134(7):509–16.

32. Fruchter R, Schrotenboer T, Luth GP. From building information model to building knowledge model. *Computing in Civil Engineering.* 2009;2009:380–9.

33. Ganah A, John GA. Integrating building information modeling and health and safety for onsite construction. *Safety and Health at Work.* 2015;6(1):39–45.

34. Zhang S, Boukamp F, Teizer J. Ontology-based semantic modeling of construction safety knowledge: towards automated safety planning for job hazard analysis (JHA). *Automation in Construction.* 2015;52:29–41.

35. Ding LY, Zhong BT, Wu S, Luo HB. Construction risk knowledge management in BIM using ontology and semantic web technology. *Safety Science.* 2016;87:202–13.

36. Kim H, Lee HS, Park M, Chung B, Hwang S. Information retrieval framework for hazard identification in construction. *Journal of Computing in Civil Engineering.* 2015;29(3).

37. Shen X, Marks E. Near-miss information visualization tool in BIM for construction safety. *Journal of Construction Engineering and Management.* 2016;142(4).

38. Zou Y, Kiviniemi A, Jones SW. Developing a tailored RBS linking to BIM for risk management of bridge projects. *Engineering, Construction and Architectural Management.* 2016;23(6):727–50.

39. Sacks R, Perlman A, Barak R. Construction safety training using immersive virtual reality. *Construction Management and Economics.* 2013;31(9):1005–17.

40. Steuer J. Defining virtual reality: dimensions determining telepresence. *Journal of Communication.* 1992;42(4):73–93.

41. Reiners D, Stricker D, Klinker G, Müller S. Augmented reality for construction tasks: doorlock assembly. *Proc IEEE and ACM IWAR*. 1998;98(1):31–46.
42. Li H, Chan G, Skitmore M. Visualizing safety assessment by integrating the use of game technology. *Automation in Construction*. 2012;22:498–505.
43. Albert A, Hallowell MR, Kleiner B, Chen A, Golparvar-Fard M. Enhancing construction hazard recognition with high-fidelity augmented virtuality. *Journal of Construction Engineering and Management*. 2014;140(7):04014024.
44. Bosché F, Abdel-Wahab M, Carozza L. Towards a mixed reality system for construction trade training. *Journal of Computing in Civil Engineering*. 2015;30(2):04015016.
45. Lu X, Davis S. How sounds influence user safety decisions in a virtual construction simulator. *Safety Science*. 2016;86:184–94.
46. Zhao D, McCoy A, Kleiner B, Feng Y. Integrating safety culture into OSH risk mitigation: a pilot study on the electrical safety. *Journal of Civil Engineering and Management*. 2016;22(6):800–7.
47. Lee G, Cho J, Ham S, Lee T, Lee G, Yun SH, et al. A BIM- and sensor-based tower crane navigation system for blind lifts. *Automation in Construction*. 2012;26:1–10.
48. Li H, Chan G, Huang T, Skitmore M, Tao TYE, Luo E, et al. Chirp-spread-spectrum-based real time location system for construction safety management: a case study. *Automation in Construction*. 2015;55:58–65.
49. Teizer J, Cheng T, Fang Y. Location tracking and data visualization technology to advance construction ironworkers' education and training in safety and productivity. *Automation in Construction*. 2013;35:53–68.
50. Zhang SJ, Teizer J, Pradhananga N, Eastman CM. Workforce location tracking to model, visualize and analyze workspace requirements in building information models for construction safety planning. *Automation in Construction*. 2015;60:74–86.
51. Zhong RY, Peng Y, Xue F, Fang J, Zou WW, Luo H, et al. Prefabricated construction enabled by the internet-of-things. *Automation in Construction*. 2017;76:59–70.
52. Lee K-P, Lee H-S, Park M, Kim H, Han S. A real-time location-based construction labor safety management system. *Journal of Civil Engineering and Management*. 2014;20(5):724–36.
53. Li N, Li S, Becerik-Gerber B, Calis G. Deployment strategies and performance evaluation of a virtual-tag-enabled indoor location sensing approach. *Journal of Computing in Civil Engineering*. 2012;26(5):574–83.
54. Cheng T, Teizer J. Real-time resource location data collection and visualization technology for construction safety and activity monitoring applications. *Automation in Construction*. 2013;34:3–15.
55. Naticchia B, Vaccarini M, Carbonari A. A monitoring system for real-time interference control on large construction sites. *Automation in Construction*. 2013;29:148–60.
56. Soleimanifar M, Shen X, Lu M, Nikolaidis I. Applying received signal strength based methods for indoor positioning and tracking in construction applications. *Canadian Journal of Civil Engineering*. 2014;41(8):703–16.
57. Cheng T, Migliaccio GC, Teizer J, Gatti UC. Data fusion of real-time location sensing and physiological status monitoring for ergonomics analysis of construction workers. *Journal of Computing in Civil Engineering*. 2013;27(3):320–35.
58. Ren W, Wu Z. Real-time anticollision system for mobile cranes during lift operations. *Journal of Computing in Civil Engineering*. 2015;29(6):04014100.
59. Teizer J, Cheng T. Proximity hazard indicator for workers-on-foot near miss interactions with construction equipment and geo-referenced hazard areas. *Automation in Construction*. 2015;60:58–73.

60. Wang J, Razavi SN. Low false alarm rate model for unsafe-proximity detection in construction. *Journal of Computing in Civil Engineering*. 2016;30(2):04015005.
61. Isaac S, Edrei T. A statistical model for dynamic safety risk control on construction sites. *Automation in Construction*. 2016;63:66–78.
62. Robson C. *Real World Research* (2nd ed.). Malden: Blackwell Publishing; 2002.
63. Seidman I. *Interviewing as Qualitative Research: A Guide for Researchers in Education and the Social Sciences*. New York: Teachers College Press; 2013.

6 Commitment, Preparation and Implementation of BIM-WHS

6.1 Case Study Investigation of the Australian Construction Industry

6.1.1 Introduction

As a case study, the Australian construction industry is used to explore the theoretical framework developed in Chapter 2. The conclusion reached in Chapter 3 is that the Australian construction industry is affected by a high rate of work-related injuries and fatalities. As a result, the case study is considered appropriate for inclusion in this research book. Using the case study, this chapter addresses the three stages of commitment, preparation, and implementation using a rigorous research methodology, which is discussed in more detail in the following section.

6.1.2 Research Approach

Creswell (1) indicated that mixed methods are the most effective for performing research in organisational and management studies. In the 1990s, the notion of using mixed methods, a combination of qualitative and quantitative methods with different typologies, was offered to spread the repertoire of health studies and social science (1–3). Giddings (4) suggested that mixed methods offer researchers a bridge between diverse types of methodologies and existing paradigms to solve complex problems in the real world. In addition, Giddings (4) argued that findings from mixed-method investigations would offer more certainty and evidence, and this would lead to more confidence in an outcome's true value. Therefore, a Qual→Quan design was selected for conducting the mixed-method investigation. This included performing an initial quality-driven study of the literature to provide and improve the results of a subsequent quantitative study (as the principal method), which was described as sequential exploratory strategy by Creswell (1). One of the main strengths of this design is its straightforward nature. This design feature facilitates describing and reporting a phenomenon as a result of having clear and separate stages (1).

DOI: 10.1201/9781003224853-6

6.1.3   *Literature Review*

The objective of the present study was to identify the key activities that are required for a prosperous adoption of BIM applications for WHS management in the industry. As described by Shields and Tajalli (5), these objectives were deemed to be in line with the capabilities of a systematic literature review approach. In other words, the core purpose of a systematic literature review is to draw upon existing research and deliver a setting for identifying, describing and converting information into a higher order of theoretical understanding and delivering the concepts, constructs and associations (6).

As illustrated in Figure 6.1, this study adopted a four-step process to carry out a systematic literature review, following the approaches taken by Pawson, Greenhalgh (7) and Chong, Lee (8).

As a result, 24 articles were found to be relevant to CSFs for BIM adoption. To ensure that all related articles are involved in the data set, Webster and Watson (9) recommended taking into account the reference lists of the collected articles, as illustrated in Figure 6.1. By taking this approach, three additional articles were identified to be relevant, bringing the overall number of articles to 27.

The literature shows that BIM-based approaches have been developed to assist WHS management during a project's lifecycle. What can be seen from these efforts is that there has been an emphasis on automated/semi-automated identification and mitigation of risk drivers as early as possible and on managing real-time risks before any occurrences of hazards.

Zou, Kiviniemi (10) indicated that despite all of the efforts made, there are some shortcomings in the current approaches. Eastman, Lee (11), and lately, Vakilinezhad, Dias (12) pointed out four main problems in automatic rule-checking systems:

- The most common rule-checking systems rely on IFC as an input data format and are currently limited in what they support.
- There is inflexibility in the rule-checking tools at the scale of all sections of a project's codes to cope with WHS risks.
- There are current efforts to enable checking the final stage of design; however, these fail to support its development process.
- There is inconsistency in the mapping terminology and vagueness regarding the narratives in the regulations.

Although 4D construction planning has been applied to safety risk planning before the actual operation, a significant shortcoming exists. By nature, the planning process is established based on knowledge and experience-based human assumptions. Because construction is a dynamic process that may last for many years and frequently involves unexpected changes and unplanned risks, operational risk management cannot normally fully comply with the original plan. Regarding this issue, an additional option is to work on a collaborative 4D construction planning platform by collecting as much reliable multidiscipline knowledge and experience

```
┌─────────────────────────────────────────────────────────────┐
│                   Step 1: Clarify the scope                  │
│                                                              │
│   Aim: Providea literaturereview of CSFs for the adoption of BIM. │
└─────────────────────────────────────────────────────────────┘
                              │
                              ▼
┌─────────────────────────────────────────────────────────────┐
│                  Step 2: Search for Evidence                 │
│                                                              │
│                      ┌──────────────┐                        │
│                      │   Keywords   │                        │
│                      └──────────────┘                        │
│                              │                               │
│                      ┌──────────────────────────┐            │
│                      │ "Web of Science" and "Scopus" │        │
│                      └──────────────────────────┘            │
│                              │                               │
│                      ┌──────────────────────────┐            │
│                      │ Title/Abstract/keyword search │        │
│                      └──────────────────────────┘            │
│                              │                               │
│             Is the paper presenting a CSF for BIM adoption?  │
│                              │                               │
│                              ▼                               │
│  ┌──────────┐   No    ◇──────────────◇                       │
│  │ Omission │◀────────│  Relevancy   │                       │
│  └──────────┘         ◇──────────────◇                       │
│                              │                               │
│                              ▼                               │
│                  ┌──────────────────────────┐                │
│                  │  Spotting relevant publications │          │
│                  │      cited in the papers  │                │
│                  └──────────────────────────┘                │
│                              │                               │
│                    No        ▼                               │
│  ┌──────────┐         ◇──────────────◇                       │
│  │ Omission │◀────────│  Relevancy   │                       │
│  └──────────┘         ◇──────────────◇                       │
│                              │                               │
│                              ▼                               │
│        ╭─────────────────────────────────────────────╮      │
│        │ Tabulating publications based on stages of the │    │
│        │           innovation adoption model.          │     │
│        ╰─────────────────────────────────────────────╯      │
└─────────────────────────────────────────────────────────────┘
                              │
                              ▼
┌─────────────────────────────────────────────────────────────┐
│         Step 3: Appraising and extracting data (Findings)    │
└─────────────────────────────────────────────────────────────┘
                              │
                              ▼
┌─────────────────────────────────────────────────────────────┐
│        Step 4: Synthesising the evidenceand discussions      │
└─────────────────────────────────────────────────────────────┘
```

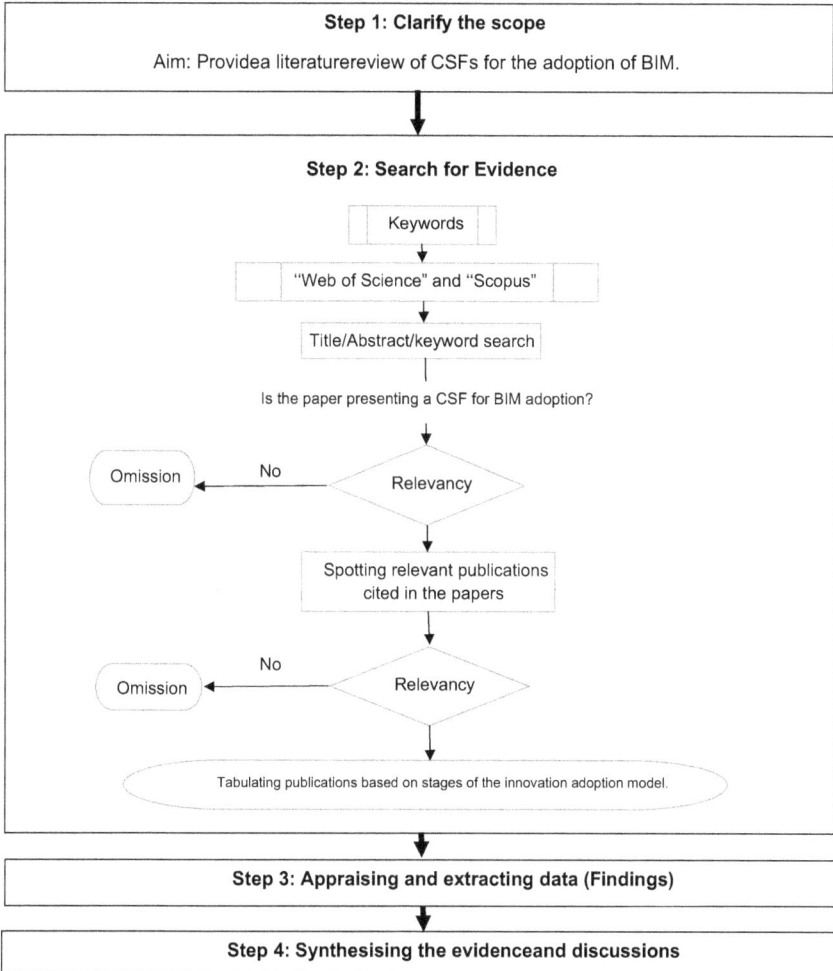

Figure 6.1 Literature review processes to categorise the literature findings

as possible (13). Another alternative approach is to use RTLS for real-time data collection and treatment. However, much of the cited work on proactive systems is still very young. Existing research places more emphasis on worker location and less on their motions and postures (which have a serious impact on construction safety), more on crane operations and less on other equipment. RTLS is also insufficiently developed to support site monitoring, which has restricted the extensive application of BIM in practice (14).

An important challenge for knowledge-management systems is how to ensure that the knowledge and experience shared by a limited number of professionals includes complete and correct information regarding the potential risks (14).

VR/AR facilitates safety training in a visual, interactive and cooperative way. However, existing research has mainly focused on development or customisation of specific approaches or platforms for one or some aspects of safety training (15, 16) and lacks a comprehensive safety training approach or platform, leading to high training costs and low efficiency. Although some studies have proposed general safety training approaches (17, 18), most have primarily focused on the benefits of visualisation, with less consideration for interaction and cooperation.

Because construction projects are one-off endeavours with numerous special features and risks existing during the whole dynamic process, any new methods for risk management are valuable when core project participants begin to use these enhanced technologies as part of their daily work (19). Ivory and Alderman (20) found that strong client leadership may suppress the sharing of ideas and create an overly narrow focus on particular types of innovation, yielding negative consequences for innovation. In addition, the projects that have been associated with the involvement of a champion to shepherd the innovation and eliminate roadblocks have experienced smooth implementation of the innovation (21). Gambatese and Hallowell (21) also identified upper management support and lessons learned about innovations as effective enablers for introducing an innovation to a construction firm.

The list of CSFs collected from the literature for the current study had to include factors that met a number of requirements. The factors had to fit in the context of WHS management and form a manageable perspective that was feasible to be further assessed by experts. Therefore, a rule-based screening approach, as suggested by Yang, Li (22), was applied in this study. In doing so, the following list of principles was taken into account:

- **Applicability:** This principle ensures the feasibility of the assessment of a factor for the current technology level and domestic policy.
- **Comprehensiveness:** The complete coverage of factors with regard to the target.
- **Significance:** From the statistical point of view, a significant factor or criterion describes a thing or idea with a value that varies from one entity to another. Hence, in the initial screening of the factors, it should come to the forefront that each selected factor carries a specific attribute and that changing its value affects the output to some extent. In other words, some minor factors with possible impacts that are not that significant could be ignored.
- **Multi-Attribute Decision Making:** This is used to achieve an acceptable number of mutually exclusive factors. According to Yoon and Hwang (23), filtering out mutually inclusive factors minimises the risk of double-counting factors, which is undesirable in the weighting stage.

As a result of literature review illustrated in Figure 6.1, 27 unique factors fully met these principles.

Table 6.1 synthesises the findings from the existing literature on CSFs for BIM adoption within the industry's current practice.

Table 6.1 Contextual factors affecting stages of innovation (BIM for WHS) diffusion in construction companies

No	Innovation adoption stage	CSF	Reference(s)	Responsible entity/ entities
1	Commitment	Employment of R&D activities to tackle existing technological barriers	(24)	Vendors and contractor
2	Commitment	Contractor and designer adaptability to amendment of the project scope and plan	(25)	Contractor and designer
3	Commitment	Strong commitment to safe project delivery from project stakeholders	(21, 26)	Client and government
4	Commitment	Clear and realistic goals	(27)	Client, contractor and designer
5	Commitment	Mutual trust	(28)	Client, contractor and designer
6	Commitment	Appropriate allocation of risks and interests	(28)	Client, contractor and designer
7	Preparation	Implementing effective communication and data exchange protocols at all levels of decision-making within the project management team	(10)	Client, contractor and designer
8	Preparation	Creating accountabilities, expectations, roles and responsibilities for the organisation	(21, 27)	Client
9	Preparation	Implementing effective health and safety protocols in the contract	(26, 29)	Client
10	Preparation	Project manager's experience and competence	(29)	Contractor
11	Preparation	Comprehensive contractor portfolio investigation to assess their previous records and their level of awareness of the BIM applications for WHS	(25)	Client
12	Preparation	Modification of the organisation's structure to facilitate and organise communication	(21)	Contractor and designer
13	Preparation	Availability of technological resources	(11, 12)	Contractor and designer
14	Preparation	Implementation of effective communication and data exchange protocols at all levels of decision-making within the project management team	(30)	Client, contractor and designer

(*Continued*)

Table 6.1 (Continued)

No	Innovation adoption stage	CSF	Reference(s)	Responsible entity/ entities
15	Preparation	Use of lessons learned in previous projects by the project management team	(21)	Client, contractor and designer
16	Preparation	Compatibility of BIM products with the WHS performance of the industry	(11, 12)	Vendors
17	Preparation	Development of new regulations for BIM models compatibility with WHS standards	(12)	Government
18	Preparation	Existence of a coordinator with WHS knowledge as a champion to lead the process of PtD	(30)	Designer
19	Preparation	Personal motivation	(27)	Contractor and designer
20	Preparation	Personal competency	(27, 31)	Contractor and designer
21	Preparation	Establishment and communication of conflict resolving strategies	(28)	Client, contractor and designer
22	Preparation	Building a strict incentive system	(28)	Client and government
23	Implementation	Incorporation of a systematic knowledge management system to organise and classify big data collected during and after the project life cycle	(10)	Contractor
24	Implementation	Enactment of the required policies to support BIM applications in the project	(25)	Government
25	Implementation	Support and cooperation of the contractor and designer in delivering a safe project	(25)	Contractor and designer
26	Implementation	Continuous monitoring and revision of BIM models during the project's execution and rectification of any WHS issues	(13)	Contractor
27	Implementation	Systematic methodologies to carefully monitor the innovation process	(28)	Contractor

6.1.4 Semi-Structured Interview

An interview is considered an insightful discussion that is undertaken through the participation of at least two individuals to extract reliable and relevant information (32). This section describes the process of data collection from the interview sessions and the analysis methods applied to extract information.

6.1.4.1 Data Collection

Interview approaches can be categorised into three types: 1) structured, 2) unstructured and 3) semi-structured interviews. Structured interviews are conducted using a prearranged list of questions; however, the unstructured form is performed freely, without any pre-specified questions. Intertwined between these two types, semi-structured approaches include several pre-designed questions to support the flow of discussion. According to Robson (32), in semi-structured interviews, potential and relevant sub-questions can be prepared to facilitate a meaningful conversation and ensure that all aspects of the topic are covered. The semi-structured interview was deemed the most suitable approach for this exploratory study because there are limited prior studies regarding the adoption process of BIM for construction WHS. As stated by Seidman (33), a semi-structured interview allows researchers to gain and discover new subjective information.

To this end, interview sessions were conducted in a short time frame, and the general flow of the interview process is illustrated in Figure 6.2. Approaching appropriate interviewees is an important factor to ensure the validity of data collected. As suggested by Bolger and Wright (34), interviewees with multiple viewpoints can be included avoid one-sided judgments and provide homogeneous opinions and a more comprehensive range of information.

6.1.4.2 Data Analysis

Following the guidelines suggested by Guest, Bunce (35) and Francis, Johnston (36), the sample size required for analysis depends on the process of deriving new

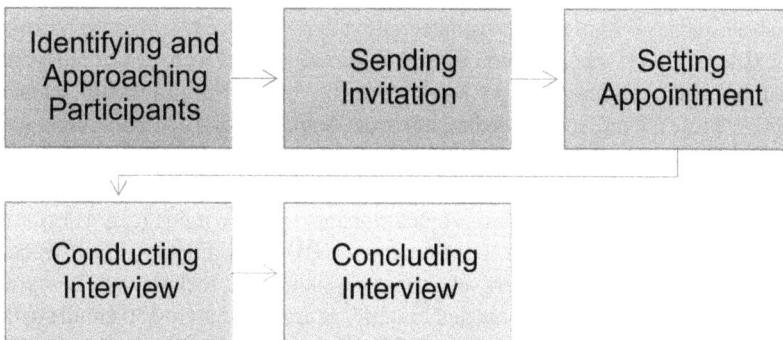

Figure 6.2 General flow of the semi-structured interviews

information and the extent of the generalisability of the findings. In this study, the interview process continued until data saturation was reached. To aid with this decision, new BIM applications and CSFs emerging from each interview were tracked, and the interviews ceased when no new codes were identified in two consecutive sessions.

The interviews were transcribed after each session and set for thematic content analysis using QSR NVivo11 software (Braun & Clarke, 2006). As explained in Figure 6.2, this is an approach used to identify, analyse and report the themes in data. This approach was deemed fit for recognising possible BIM applications regarding the control of critical WHS areas and CSFs that could facilitate the implementation of these approaches in the construction industry of Australia.

In the first step, the researcher carefully read the transcribed interviews three times to become familiar with the data. Next, initial ideas were developed considering the two main topics of the study. For the first part of the interviews, the researcher searched for applications of BIM that experts believed were effective in improving the identified critical WHS domains. For the second part of the interviews, the researcher looked for actions described by the experts as critical for the adoption of BIM for WHS purposes in three areas: Commitment, preparation and implementation. Thematic content analysis was conducted immediately after the interview sessions to check data saturation.

6.1.5   Questionnaire Survey

An online survey method was selected to validate the findings from the interview approach. Surveys are considered a means to collect information from a particular population. Due to the non-feasibility of data collection from the entire population, a survey approach facilitates data collection from a proper sample of the whole population. An appropriate selection of the sample allows for the generalisation of findings to the whole population (37).

6.1.5.1   Data Collection

In order to reach a suitable pool of participants for the questionnaire survey, a list of major companies within the Australian context that employed BIM in their projects was extracted. This was achieved by searching the official websites of BIM-based consultant and contractor companies in Australia. In addition, LinkedIn, as one of the main professional social media sites, was searched for BIM and construction safety groups within Australia, and this led to identifying most of the active actors within the industry. The suitable target group for the survey was deemed to be a range of professionals, including project managers, safety managers, safety officers, site supervisors, architects, BIM managers, BIM coordinators, civil/structural engineers, mechanical engineers, electrical engineers and technicians. The potential participants were then approached by sending approximately 550 emails to their companies or contacting them through LinkedIn's messaging application to fill out the Key Survey online questionnaire form developed based on the findings of the interview analysis. Along with the invitation, the potential participant received a participant information sheet.

6.1.5.2 Design of the Questionnaire Survey

The questionnaire survey was designed to measure three latent variables of the research's conceptual model through their relevant observable variables. The first part described the overarching aim of the study to assist respondents in acquiring an accurate understanding of the purpose of the study; this part also included several questions regarding demographics to ensure a decent sample was approached.

The second part of the questionnaire included evaluating the importance of the CSFs identified from the interviews for three latent variables: Commitment, preparation (at the organisation and project levels) and implementation (at the organisation and project levels). The respondents were asked to state their level of agreement regarding the CSFs' effects on the successful integration of BIM with the current practice to improve WHS performance (observable variables) (Table 6.2). In the third part of the questionnaire, respondents were asked to state their level of agreement with a set of questions. To elicit agreement levels, a five-point Likert scale was used (5 = strongly agree, 4 = agree, 3 = neutral, 2 = disagree, 1 = strongly disagree).

6.1.5.3 Analysis

From the received responses, the respondents' demographics were analysed based on the proportion of each group to understand whether the respondents had adequate knowledge regarding the target, a variety of rules and sufficient years of experience.

According to Ho (38), multivariate regressions represent an appropriate method of data analysis when the goal of a study is to look for links between variables and determine the strength of such links. For both exploratory and confirmatory research questions, structural equation modelling (SEM) is an effective approach (38). Hair Jr, Hult (39) categorised SEM methods into two broad approaches, partial least square (PLS) and covariance-based methods. The selection of an appropriate approach depends on the nature of the received data and the objectives devised for the study. Given that the practice of BIM in construction WHS is relatively new for many companies in Australia, finding the over 200 respondents required for covariance-based-SEM analysis (40) was not feasible, while PLS-SEM requires a much lower minimum data number. PLS-SEM was deemed to be the most appropriate method for this study due to the novelty of the conceptual model and the ability to analyse and explore links among a number of constructs using PLS-SEM (29). As extensively introduced by Hair Jr, Hult (39), SmartPLS v.3.2.1 was used as the main tool to run the analysis. Models in SEM are comprised of two main categories of variables: Observable, manifest variables, such as those measured through the questionnaire and latent variables showing the underlying constructs associated with manifest variables (38). The associations between manifest variables and the

Table 6.2 Questions in the third part of the online questionnaire survey

No.	Topic
1	Commitment of the designers to develop BIM models with possible safety features
2	Mutual trust among the parties involved in the projects to circulate required safety information in their BIM models
3	Workers' agreement to storing and using their health and safety records
4	The commitment of contractors to use BIM models in their safety management system
5	Lowering the cost of technology by vendors makes them usable for small and medium-sized companies
6	Government mandating for the use of BIM models for the safety management of the projects
7	Mandating the use of BIM models for safety in client contracts
8	The existence of financial incentives in the contracts for using BIM in the safety management process
9	Technical and Further Education (TAFE) and university initiatives for training safety officers/managers to use BIM
10	Initiatives from the Master Builders Association by introducing possible approaches to the companies
11	Vendors' initiatives in developing software that could identify unsafe designs according to national safety regulations
12	Availability of the technical hardware and software in the companies
13	Creating new roles and responsibilities within the organisations for facilitating the implementation of the BIM for safety management
14	Employing safety managers with BIM knowledge in the companies
15	Setting out clear safety goals in the contracts for using BIM in the safety management process
16	Setting the scope of the project's safety management for the entire life cycle of the projects from the first drafts
17	Allocation of additional time and budget for safety in the design process
18	Mandating the use of information provided in the BIM models for maintenance trade workers
19	Initiatives from large companies as front-liners in the industry for using BIM for safety
20	Liability of data inserted in the models for each of the parties involved
21	Continuous monitoring and modification of the process of using BIM for safety management
22	Having an effective communication and data exchange system in place among the people involved in the safety management process of the projects
23	Sufficient compatibility among the selected software packages to transfer safety features from one to another
24	Developing Intellectual Property (IP) conflict-resolving protocols to facilitate the exchange of models between designers and contractors
25	Continuous collaboration of the designers and contractors in monitoring and modifying the models during the execution of projects
26	Frequent updating of the safety information in the BIM models
27	Sub-contractors' experience and competency in using BIM
28	Engaging sub-contractors in the safety management process and providing them with required information through models

underlying constructs could be specified in formative models where it is assumed that indicators cause the constructs, as each one of the indicators captures one of the aspects of the construct. Taken jointly, the indicators determine the meaning of the construct. As a result, the breadth and comprehensiveness of the indicator's domain are central to ensuring that a construct is adequately covered and that all aspects are captured (39).

Following the submission of data to SmartPLS, several requirements about the data and the specified model should be met in order to ensure that the results of formative models are reliable. The highest priority should be given to the assessment of collinearity, which means that two or more formative indicators in a block capture exactly the same information in them. A recommended measure to evaluate collinearity is the variance inflation factor (VIF), which is calculated according to Equation 6.1, where x represents variables utilised as formative indicators. To calculate R_x^2, the indicator x is taken and regressed for all the remaining indicators of the same block (39). R_x^2 is the proportion of variance of x associated with other indicators.

$$VIP_x = \frac{1}{1-R_{x^2}} \tag{6.1}$$

In cases where the level of collinearity for formative indicators of a construct is very high (VIF $\geq$ 5), the variable should be removed from the model prior to conducting any further analysis.

In a case where no critical levels of collinearity are observed in the model, SEM-PLS analysis should be performed to analyse the significance of outer weights and interpret the formative indicator's relative and absolute contribution to the underlying constructs. Researchers should test whether the outer weights calculated in formative models are significantly different from 0 using bootstrapping. To this end, a bootstrapping function in SmartPLS was utilised with the algorithm option of no change sign and 5,000 bootstrap subsamples as a conservative configuration to calculate the significance of outer weights (39). The critical value for the significance level of 5% ($\propto$ = 0.05) was 1.96 (39).

Once the contributions of the observable variables to the latent variables were calculated, the next stage was to assess the predictive capability and association among the latent variables of the model. Since there is no goodness of fit model for PLS-SEM, Hair Jr, Hult (39) suggested evaluating the performance of a PLS-SEM model based on its ability to predict the latent variables. Therefore, the value of R^2 and the effect size f^2 for formative models were calculated to evaluate the collinearity among the latent variables and the significance of path co-efficiencies.

6.1.6 Results

6.1.6.1 Interview Results

The interviewees were selected based on their positions and experience in the industry. It was essential to ensure that the participants could provide realistic and industry-oriented information based on their experiences in the industry. There-fore, the positions were set to find a variety of experts who used BIM in their daily practice from both consultant and contractor groups. These experts included a pro-ject manager, two BIM managers, a BIM coordinator, a national HSE manager and an architect. The sample was deemed representative of a rich variety of expertise. The sample size also met the criterion set in the research design, which was no emergence of new codes in the last two interviews.

Interviewees were asked to express the critical actions required to be taken by the client, contractor, designer, software vendors and the government. The CSFs identified from the interviews were categorised into commitments, organisational preparations, project preparations, organisational implementation and project implementation factors to be used in the questionnaire survey (Table 6.2).

Although the CSFs identified were not quantitatively ranked by the interview-ees, some factors were described by the interviewees as very important for a suc-cessful adoption process. Most of the interviewees suggested that technology vendors should develop platforms that could directly address safety issues in the models, where at the current stage, WHS information is entered into the models through customisation actions taken by the construction companies. It was also suggested that technology vendors should reduce the prices of their products up to the point that the BIM software could be more attainable for smaller companies.

Interviewees described contractors as the main drivers of the innovation adop-tion process. This was because they were liable for construction accidents and, at the same time, could benefit from improving construction WHS. The adoption of BIM applications in projects for WHS purposes requires having people operate the technology and process. Interviewee 5 explicitly pointed out that "it is of impor-tance to have site safety managers who have hands-on skills to use BIM tools. Also, this requires sub-contractors to be educated to use such tools." Interviewee 6 argued that "the construction projects vary one from another, and it is hard to find a fixed approach to implement these processes, and it requires continuous monitor-ing of the process and revising it."

Interviewees also discussed the role of the clients and mentioned that they should look at the projects as lifetime assets that require time-to-time maintenance activities. Therefore, spending some money upfront to develop the building models capable of storing WHS information in the models could save money during the maintenance phase by reducing the number of accidents.

Interviewee 3 suggested that "[the] government, which is also one of the major clients, can accelerate setting the BIM plans for construction WHS. Currently, the government has planned to implement BIM for WHS by 2023, and this plan only targets large-size projects".

Interviewees who worked for contractors argued that the designers are responsi-ble for developing building models, and most of the developed BIM models at the current stage do not have a mechanism to store WHS information, which makes the contractors' job more difficult with regards to modifying the model to be capa-ble of safety management. On the other hand, designers argued that they could not be held accountable for a contractor's unsafe performance. They discussed the importance of safety in design meetings and mentioned that BIM could significantly improve the efficiency of those meetings. However, this would require the allocation of additional time and effort from both parties, the consultants and the contractors. Table 6.3 presents the CSFs to leverage BIM applications in the WHS practice of the Australian construction industry, as extracted from interviews with construction experts.

Table 6.3 CSFs for the adoption of BIM WHS applications in construction projects

	Code	*Description*
Commitment	Com1	Commitment of the designers to develop BIM models with possible safety features
	Com2	Mutual trust among the parties involved in the projects to circulate required safety information in their BIM models
	Com3	Workers' agreement to store and use their health and safety records
	Com4	The commitment of contractors to use BIM models in their safety management system
	Com5	Lowering the cost of technology by vendors to make them usable for small and medium-sized companies
	Com6	Government mandating the use of BIM models for safety management of the projects
	Com7	Mandating the use of BIM models for safety in the contracts by the clients
	Com8	Existence of financial incentives in the contracts for using BIM in the safety management process
Organisational Preparation	Prep_org1	TAFE and university initiatives in training safety officers/managers to use BIM
	Prep_org2	Initiatives from the Master Builders Association by introducing possible approaches for the companies to use
	Prep_org3	Vendor initiatives in developing software that can identify unsafe designs according to the national safety regulations
	Prep_org4	Availability of the technical hardware and software in the companies
	Prep_org5	Creating new roles and responsibilities within the organisations for facilitating the implementation of BIM for safety management
	Prep_org6	Employing safety managers with BIM knowledge in construction companies
Project Preparation	Prep_pro1	Setting out clear safety goals in the contracts for using BIM in the safety management process
	Prep_pro2	Setting the scope of the project's safety management for the entire life cycle of the project from the first draft
	Prep_pro3	Allocation of additional time and budget for safety in the design process
	Prep_pro4	Mandating the use of information provided in the BIM models for maintenance trade workers

(Continued)

Table 6.3 (Continued)

	Code	Description
Organisational Implementation	Imp_org1	Initiatives from large companies as the front-liners in the industry in using BIM for safety
	Imp_org2	Liability of data inserted in the models for each of the parties involved
	Imp_org3	Continuous monitoring and modification of the process of using BIM for safety management
	Imp_org4	Having an effective communication and data exchange system in place among the people involved in the safety management process of the projects
Project Implementation	Imp_pro1	Sufficient compatibility among the selected software packages to transfer safety features from one to another
	Imp_pro2	Developing IP conflict-resolving protocols to facilitate the exchange of models between designers and contractors
	Imp_pro3	Continuous collaboration of the designers and contractors for monitoring and modifying the models during the execution of projects
	Imp_pro4	Frequent updating of the safety information in the BIM models
	Imp_pro5	Sub-contractors' experience and competency in using BIM
	Imp_pro6	Engaging sub-contractors in the safety management process and providing them with required information through models

6.1.6.2 Questionnaire Results

The survey was launched in March 2018, and 176 experts participated in this survey from across Australia. Another 153 participants began the online survey but did not submit, as the records of the Key Survey show. The response rate for this survey was 53.50%, considering the incomplete surveys. According to Hair Jr, Hult (39), G*Power software is a powerful tool for measuring the required sample size for a PLS-SEM analysis. According to the findings of the interviews, the highest number of predictors was 17, which belonged to the evaluation of BIM stage. Using a post-hoc function of the G*Power software, a minimum population effect size of 74 was required (Figure 6.3), and the collected sample size was much larger.

As stated in the research design, the first step in conducting a PLS-SEM analysis to evaluate collinearity was to calculate VIF using SPSS software, calculated according to Equation 6.1.

The VIF ranges were as follows: Commitment (1.19–1.58), project preparation (1.04–1.06), organisation preparation (1.03–1.33), project implementation (1.11–1.47) and organisation implementation (1.06–1.23). These values were found to be significantly lower than the acceptable range of VIF $\leq$ 5 and confirmed that there was no collinearity issue.

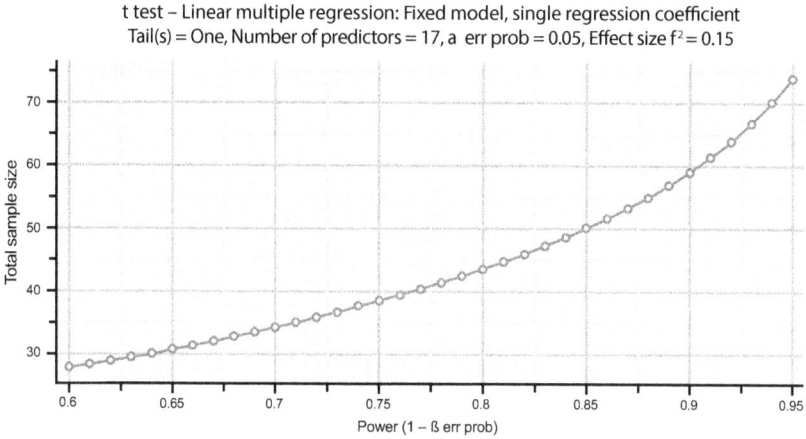

Figure 6.3 Minimum population size required for the PLS-SEM analysis

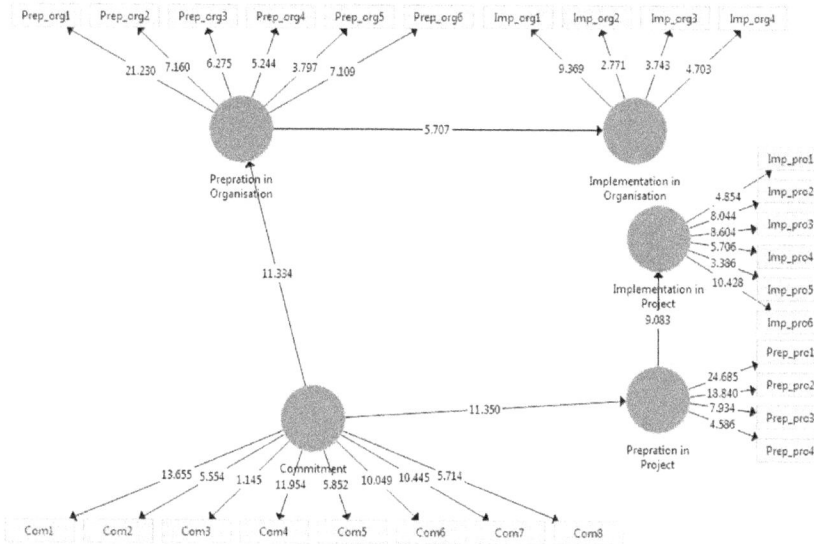

Figure 6.4 Initial calculation of t-values for outer weights based on 5,000 rounds of sub-sample bootstrapping

Figure 6.4 presents the results of running this analysis by setting the algorithm option to "no change sign" and allowing 5,000 bootstrap sub-samples as a conservative configuration.

As illustrated in the figure, apart from that of one of the observable variables (Com3), all VIF values were higher than 1.96, which shows that these variables could remain in the model as they significantly contributed to the latent variables.

However, for the Com3 factor, the value of the outer loading was used to indicate whether this factor could remain or should be removed. With reference to Hair Jr, Hult (39), factors can only remain if their outer loading value is higher than 0.5. In the case of Com3, the outer loading value was 0.145, which indicated that this factor should be removed from the list.

Because one of the factors was removed from the list, a second run of the bootstrapping function was performed to ensure that all variables possessed an acceptable t-value. Figure 6.5 presents the results of the second run and shows that all variables had t-values higher than 1.96.

After this modification, the model was considered fit to develop the path coefficients and determine the relative contribution of the observable variables to their relative constructs. Figure 6.6 presents the findings of running SmartPLS software and shows which variables should be focused on for a successful adoption of BIM to reduce occupational fatalities.

In the commitment stage, the most critical factors were identified as the commitment of the designers to develop BIM models with possible safety features (Com1) and the commitment of contractors to use BIM models in their safety management system (Com4). These factors were followed by mandating approaches imposed by the clients and government (Com7 and Com6).

In the organisational preparation stage, two factors were deemed to be particularly effective: TAFE and university initiatives in training safety officers/managers to use BIM (Prep_org1) and employing safety managers with BIM knowledge in construction companies (Prep_org6). In addition, two project preparations were

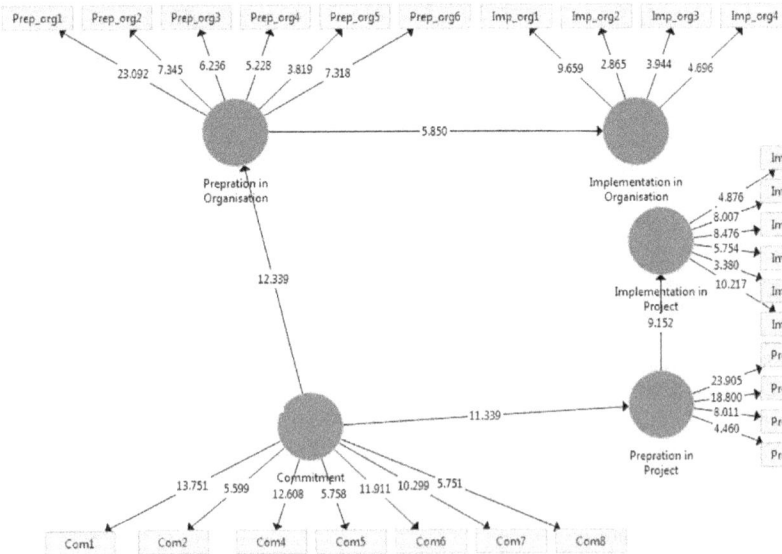

Figure 6.5 The second calculation of t-values for outer weights based on 5,000 rounds of sub-sample bootstrapping

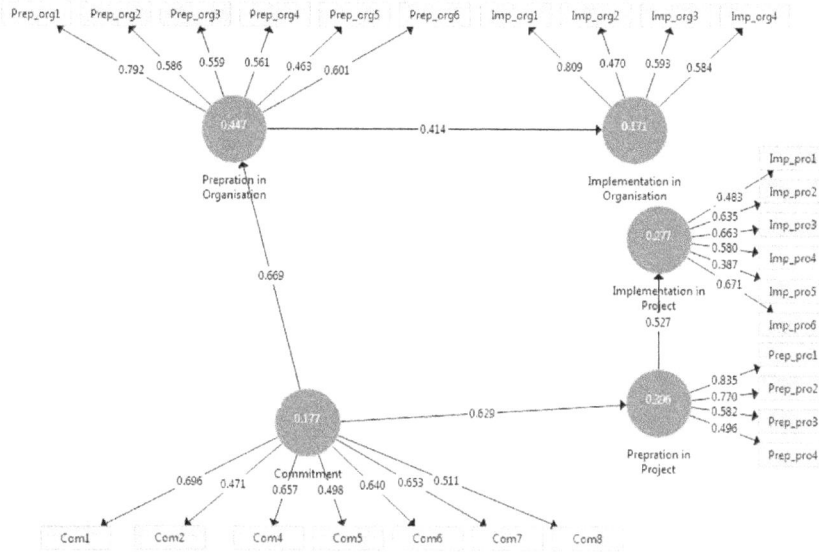

Figure 6.6 Final PLS-SEM model presenting path co-efficiencies and relative contributions of the observing variables to the constructs

determined to have significantly higher contributions than others: Setting out clear safety goals in the contracts for using BIM in the safety management process (Prep_pro1) and setting the scope of the project's safety management for the entire life cycle of the project from the first draft (Prep_pro2).

As illustrated in Figure 6.6, significant weight in the organisational implementation was given to the critical factors of initiatives from large companies as the front-liners in the industry in using BIM for safety (Imp_org1). The next two most significant factors in this construct were the continuous collaboration of the designers and contractors in monitoring and modifying the models during the execution of projects (Imp_org3) and the existence of an effective communication and data exchange system in place among the people involved in the safety management process of the projects (Imp_org4).

Critical factors in the project implementation construct were identified as engaging sub-contractors in the safety management process and providing them with required information through models (Imp_pro6). The next most significant factor in this construct was the continuous collaboration of the designers and contractors in monitoring and modifying the models during the execution of projects (Imp_pro3). R^2 values were replaced in Equation 6.1 to calculate the VIF of the latent constructs. Table 6.4 shows that all VIF values were under the threshold of 5.

Regarding the impact of the predictor constructs, (39) suggested that using the values of affect size values lower than 0.02 presents a minor impact. Table 6.5 presents the f^2 values of the latent variables, and there was no f^2 value below 0.02,

Table 6.4 R square and VIF values of the latent constructs of the model

Latent constructs	R^2	VIF
Commitment	0.177	1.215
Implementation at the organisation level	0.171	1.206
Implementation at the project level	0.277	1.383
Preparation at the organisation level	0.447	1.808
Preparation at the project level	0.396	1.656

Table 6.5 f^2 values for the structural model

	1	2	3	4	5
1: Commitment				0.81	0.655
2: Implementation at the organisation level					
3: Implementation at the project level					
4: Preparation at the organisation level		0.206			
5: Preparation at the project level			0.384		

which shows acceptable model fitness. Thus, the conceptual framework of the research presented in the Chapter 2 was supported, and the latent variables were positively related.

A discussion of the results of the case study is presented in the next section.

6.2   Role of Clients and Governments

As demonstrated in Table 6.6, the findings of this study show the critical role of clients among the commitments required for the successful adoption of BIM for construction safety management. Previous studies have also confirmed these findings, as clients represent the funding source of the projects, and the adoption of innovative ideas strongly requires their support (41). Contractors are less likely to employ BIM for WHS management if the clients do not support it. The existence of financial incentives in contracts to encourage the use of BIM in the safety management process was identified as a valuable commitment to be made by clients. The more experienced and demanding the client, the more likely they are to support the integration of innovations into the contracts (42).

As the biggest client in the country, the Australian government can have a significant influence by mandating the use of BIM for construction WHS management in public projects. Tam, Zeng (43) highlighted the critical role of governments in enforcing that companies must improve their WHS performance. Blayse and Manley (44) argued that the process of developing regulations that suit the adoption of a new technology is a complex process that mainly depends on the existence of sufficient knowledge among the industry's key players and the development of appropriate mechanisms. One of the best examples of such enforcement is the UK's enforcement of the use of BIM in projects (45). In the latest update of the

Table 6.6 The most effective CSFs to be considered by the responsible entities

Stage	CSF	Responsible entity/entities
Commitment	Commitment to developing BIM models with possible safety features	Designers
	Commitment to using BIM models in the safety management of the projects	Contractors
	Mandating the use of BIM models for the safety management of the projects	Government, clients
	The existence of financial incentives in contracts for the use of BIM in the safety management process	Clients
Preparation at the organisation level	Training safety officers/managers to use BIM	Educational bodies
	Employing safety managers with BIM knowledge	Contractors
	Introducing potential BIM applications for WHS management to the companies	Professional bodies
	Availability of technical hardware and software	Designers, contractors
	Developing software to identify unsafe designs according to national safety regulations	Vendors
Preparation at the project level	Setting out clear safety goals for using BIM in the safety management process	Contractors
	Setting the scope of the project's safety management for the entire life cycle of the project from the first drafts	Clients
	Allocation of additional time and budget for safety in the design process	Designers
Implementation at the organisation level	Initiatives from large companies as the front-liners in the industry in using BIM for safety	Contractors
	Continuous monitoring and modifying the process of using BIM for safety management	Contractors
	Having an effective communication and data exchange system in place among the people involved in the safety management process	Contractors
Implementation at the project level	Engaging sub-contractors in the safety management process and providing them with required information through models	Contractors
	Continuous collaboration of designers and contractors in monitoring and modifying the models during the execution of the projects	Designers, contractors
	Developing IP conflict-resolving protocols to facilitate the exchange of models between designers and contractors	Designers, contractors
	Frequent updating of the safety information in the BIM models	Contractors

BIM standards, the British Standards Institute (2018) introduced BSI: PAS 1192–6 (46), which specifies collaborative sharing and the use of structured health and safety information using BIM. This standard highlights the process of integrating traditional risk management systems with BIM for use in the project life cycle. As identified in this study, it is important that governments and clients set the use of BIM for the entire life cycle of the project, as many accidents take place during the maintenance phase of projects. Having access to WHS information and having a structured risk management plan in place can prevent most fatalities.

6.3   Role of Contractors

Contractor companies hold most of the liability when a construction accident occurs. They must therefore innovate to improve their WHS performance. Blayse and Manley (44) highlighted the importance of diffusing new technologies as the main criterion for contractor companies to remain competitive in the industry and present their improvements in operations and distinctive technical capabilities. As identified in this study, the process of BIM adoption for WHS management by the industry is highly dependent on the contractors' commitment.

Safety managers in construction companies hold the prime responsibility of WHS management. Blayse and Manley (44) noted that some in-house technical competence is necessary to benefit from and absorb new technologies fully. Gann (47) argued that absorptive capacity is more important than technical capability and prior knowledge functions. Therefore, it is important for contractors to employ a safety manager with BIM knowledge to be able to interpret and act upon the WHS management of the projects. Winch (48) stated,

> Innovations need champions. Ideas are carried by people, and ideas are the rallying point around which collective action mobilises. Unless the 'systems integrator' is convinced of the merits of the new idea, and has the skills to incorporate it into the system as a whole, change is likely to be slow.
>
> (p. 274)

Safety managers with BIM knowledge can be considered the champions for a successful adoption process. Nam and Tatum (49) noted that a champion's technical competence enables overcoming the uncertainties involved in the adoption of new technologies, while power enables challenging the resistance to innovation.

Blayse and Manley (44) indicated that a limited number of construction companies have the capacity to develop a R&D programme. This places the burden on large companies that have the capability to initiate R&D programmes to devise an effective implementation process. In addition to hardware capabilities, contractor companies are required to enhance their software capabilities. A well-structured BIM model would be of no use to companies that do not have a license to use it.

Banihashemi, Hosseini (29) indicated the critical role of having an appropriate communication and information exchange system in place in the integration of Information and Communications Technology (ICT) for construction companies.

WHS data are required to be handled by the right people and responded to in a short time. Although utilisation of BIM structures the data exchange process, this requires proper management of the process. The current study also recognises the importance of the time-to-time modification of the process and updating models at the implementation stage. Integration of BIM with the current WHS management of construction projects was deemed to be a new approach and required learning from previous experiences and modifying the process. Gann (47) noted that contractor companies often have difficulty learning from previous projects. Blayse and Manley (44) suggested codifying the knowledge learned from projects so that it can be easier to diffuse into future projects.

Blayse and Manley (44) suggested that imposing strict high standards in projects can force contractors to adopt new technologies. However, Gann, Wang (50) noted that setting complex standards for projects discourages innovation adoption, and simplicity and clarity are required to enable the diffusion of good practices and encourage innovation. The current study found that setting clear safety goals for using BIM in the safety management process is a critical factor in organisation preparation.

The collaboration between contractors, sub-contractors and the design team in the implementation stage was a significant factor in the adoption process. Construction projects usually involve many sub-contractors who undertake most of the project's activities together. The main contractor's responsibility is to manage them to perform their tasks efficiently and safely. As identified in the previous section, the risk management process is highly in need of sub-contractors who are experienced in using BIM to engage in the process. Another key collaboration is between the contractor and the design team. Gambatese, Behm (51) noted the insufficient level of collaboration among the design team and contractor in the safety-in-design meetings. Gu and London (52) indicated that ownership of intellectual property (IP) and protection of copyrights is a major conflict between designers and contractors. Both groups often prefer to own the model, which might be reused to another company's advantage.

6.4 Role of the Design Team

Whether it is called "prevention through design" or "safety in design", the process highlights the high commitment of the design team in devising structures that are safe to construct. Schulte, Rinehart (53) defined this process as:

> The practice of anticipating and designing out potential occupational safety and health hazards and risks associated with new processes, structures, equipment, or tools, and organising work, such that it takes into consideration the construction, maintenance, decommissioning, and disposal/recycling of waste material, and recognising the business and social benefits of doing so.
>
> (p. 115)

A study by Tymvios and Gambatese (54) noted that architects have a much lower willingness to commit to safety in the design process than the client and contractor.

The main barriers to the design team are identified as legal, economic and contractual obstacles (54). The current study also identified the requirement for time and budget allocation as a CSF in the project preparation stage.

Although design teams might have sufficient knowledge of BIM, they often lack safety knowledge. A study by Tymvios, Gambatese (30) argued that most design team members do not have knowledge about safety in design. There are currently several BIM-based tools, such as the Solibri Model Checker, that can assist designers in applying safety rules into their designs. As a critical preparation, design teams require a good arrangement of hardware and software to make this possible.

6.5 Role of the Vendors and Technology Providers

Vendors and technology providers play a significant role in the establishment of BIM applications for construction WHS. The technology here is very new for most construction companies, and this requires the development of more user-friendly platforms. As such, Eastman, Lee (11) and Vakilinezhad, Dias (12) argued that the current tools do not address WHS issues and are more focused on design aspects, as well as the time and cost management of the projects. Tools such as Autodesk Navisworks, Asta PowerProject and Syncro 4D construction project management software and 3D BIM software allow visually detection of WHS concerns, while WHS management processes require more use of embedded information. This problem has been addressed in research projects. For example, Arslan, Riaz (55) integrated environmental sensors with BIM models to detect unsafe environmental conditions. Bahn (56), Benjaoran and Bhokha (57), Choe and Leite (58) and Dong, Wang (59) integrated traditional risk management approaches with 4D BIM tools to create accessible database of WHS management information. Additionally, some research has focused on the development of semi-automated/automated site design tools that consider the unsafe proximity of the workers to heavy equipment (60–62).

Software such as the Solibri Model Checker represents a more advanced approach that can identify unsafe designs in a semi-automated/automated way. However, the current state of these tools is limited. Vakilinezhad, Dias (12) described these limitations as highly dependent on what the IFC supports, inflexibility in rule-checking tools and inability to detect WHS issues during the initial phases of design. In the case of Australia, there is a lack of such tools that can check the models against the national WHS regulations.

6.6 Role of the Professional and Educational Bodies

For the Australian construction industry, Hardie and Newell (63) noted the professional bodies' effectiveness in the adoption of new technologies. The results of the current study showed that professional bodies can support joint problem solving that can encourage the sharing of tacit knowledge related to BIM's WHS potential. The intervention of such organisations can reduce the risk of adoption, which can lead to higher consumption of the technology. On the other hand,

Blayse and Manley (44) argued that construction companies are discouraged from adopting innovative approaches when professional bodies introduce inflexible guidelines.

Educational bodies such as universities and TAFE institutions have recently started to provide BIM courses, with this now becoming a more common practice. However, the training of project/safety managers to incorporate BIM into their WHS management practices is missing in the current stage. Although there are some initiatives from universities, such as the University of Newcastle, which has two courses on BIM-enabled safety management for construction management students at the bachelor level, the units are more about using the visual aspect of BIM models rather than the embedded information.

6.7 Implementation Framework of BIM for Construction WHS

This section describes the overarching aim of the study, which was to map the process of diffusing BIM in construction WHS management (Figure 6.7). This process began with the identification of the critical causations of fatal accidents within Australian construction projects. The factors of risk management, workers' actions and behaviours, worker's capabilities, immediate supervision, temporary and permanent work design, construction process, workers' health and fatigue and equipment condition were found to be the most critical causations. This recognition was achieved by drawing accident causation networks based on the ConAC model for each of the accident cases collected from the NCIS database.

The second phase of innovation adoption model development was to evaluate BIM applications regarding causations. It was determined that risk causation management could be controlled using a suitability assessment of the dynamic equipment/machinery via 4D BIM simulations; planning ahead for the required materials and equipment for safer execution of the projects using 4D BIM planning; communication of safety issues through software such as BIM 360 Field, which circulates information through tablets and smartphones and informs people that are responsible for taking actions; and using BIM as a digital library of the site condition. BIM can support workers' actions and behaviours by testing and improving workers' safety culture in situations such as reporting unsafe conditions on the site, recognising the required PPE, communicating safety concerns with their colleagues, etc., and using BIM as a digital library of workers' safety track records. Workers' WHS capabilities can be improved by training workers in an interactive virtual environment, which can involve all existing risks, unlike traditional classroom-based training, and assessment of workers and sub-contractors' capabilities by conducting a walkthrough of the 4D simulations and discussing their approach to safely performing their tasks.

Immediate supervision at construction sites can be further automated by monitoring workers and equipment using attached location tracking sensors that are connected to a BIM model and controlled by the project management team. Other possibilities include frequent monitoring of the site and equipment conditions using flying drones and tracking of environmental conditions such as temperature, humidity, wind speed, etc., in the indoor and outdoor locations through sensors that are linked to a BIM model.

BIM can improve temporary and permanent work designs through automated or semi-automated checking of the building designs in BIM models to avoid unsafe designs or identify possible hazards as early as possible. Further option includes automated or semi-automated designing, checking the design of on-site temporary structures using validated algorithms to avoid design related failures and using BIM models provided by suppliers to assess the suitability of the materials and equipment they provide for the construction site.

The construction process can be made safer by developing safe work method statements for companies and trade unions. BIM can also support workers by monitoring their health and fatigue conditions using remote health monitoring sensors that can capture the location and health condition of workers and show them in a BIM model, using BIM to develop a digital library of the workers' health. Equipment conditions can be monitored through using BIM as a digital library of the equipment/machinery maintenance information. An external database can be created to store operation and maintenance information on the equipment/machinery, and this can be linked to the relevant equipment/machinery used in the 4D BIM models.

The commitments identified in this study for successful diffusion of an innovation were: Designer commitment to developing BIM models with possible safety features, contractor commitment to using BIM models in the safety management of the projects, government and clients mandating the use of BIM models for safety management of the project and client provision of the financial incentives in the contracts for using BIM in the safety management process.

Two types of preparations were deemed to be required: Organisational- and project-level preparations. Educational bodies must train safety/project managers to use BIM for the WHS management of projects. Meanwhile, contractors should employ safety managers with knowledge of BIM. This also requires preparations by professional bodies such as the Australian Institute of Building to promote BIM applications for construction WHS management. Design and contractor teams must prepare suitable software and hardware that can support the implementation of BIM-based WHS management. Additionally, technology providers should adjust their products to an extent that can support WHS management in the design, construction and maintenance of the project.

During projects, contractors should set clear safety goals for using BIM in the safety management process, and clients should set the scope of the project's safety management for the entire life cycle of the project. In addition, the design team should allocate additional time and budget for safety in the design process.

As in the preparation stage, implementations also fall into categories of project-level and organisational. Large companies should begin the implementation of BIM for WHS management, as they have the required resources. Where an innovation is adopted by a company, the process of implementation requires monitoring over time. Organisations should also apply effective communication and data exchange systems among the people involved in the safety management process.

Identified critical accident causes → **Evaluation of BIM potentials to intervene** → **Critical commitments** → **Critical preparations** → **Critical implementations**

Identified critical accident causes:
- Risk management
- Workers actions and behaviours
- Worker capabilities
- Immediate supervision
- Work design
- Construction process
- Workers health and fatigue
- Equipment condition

Evaluation of BIM potentials to intervene:
- 4D risk assessment at planning stage
- Virtual training, education and method statement development
- Project monitoring and management at construction stage using visualised sensing technologies
- Digital knowledge management of the project OHS information
- Prevention through permanent and temporary work design

Critical commitments:
- Commitment of the designers to develop BIM models with possible safety features
- Commitment of contractors to use BIM models in their safety management process
- Government mandating the use of BIM models for safety management in the projects
- Mandating the use of BIM models for safety in the contracts by the clients
- Lowering the cost of technology by vendors
- Existence of the financial incentives in the contracts
- Mutual trust among the parties involved in the projects to circulate required safety information in their BIM models

Critical preparations:
- Educational system's initiatives in training safety officers/managers to use BIM
- Initiatives from the professional bodies to introduce and promote potential approaches
- Companies employment of safety managers with BIM knowledge *(Organisational)*
- Vendors' initiatives in developing safety specific software
- Availability of the technical hardware and software in the companies
- Creating new roles and responsibilities within the organisations
- Setting out clear safety goals in the contracts
- Setting scope of the projects safety management for the entire life-cycle of the projects *(Project)*
- Allocation of additional time and budget for safety in the design process
- Mandating the use of information provided in the BIM models for the maintenance trade workers

Critical implementations:
- Initiatives from large companies in using BIM for safety
- Having in place an effective communication and data exchange system
- Continuous monitoring and modifying the process *(Organisational)*
- Liability of data inserted in the models
- Engaging subcontractors in the safety management process
- Continuous collaboration of the designers and contractors
- Developing IP conflict resolving protocols
- Frequent updating of the safety information in the BIM models *(Project)*
- Sufficient compatibility among the selected software packages
- Sub-contractors' experience and competency in using BIM

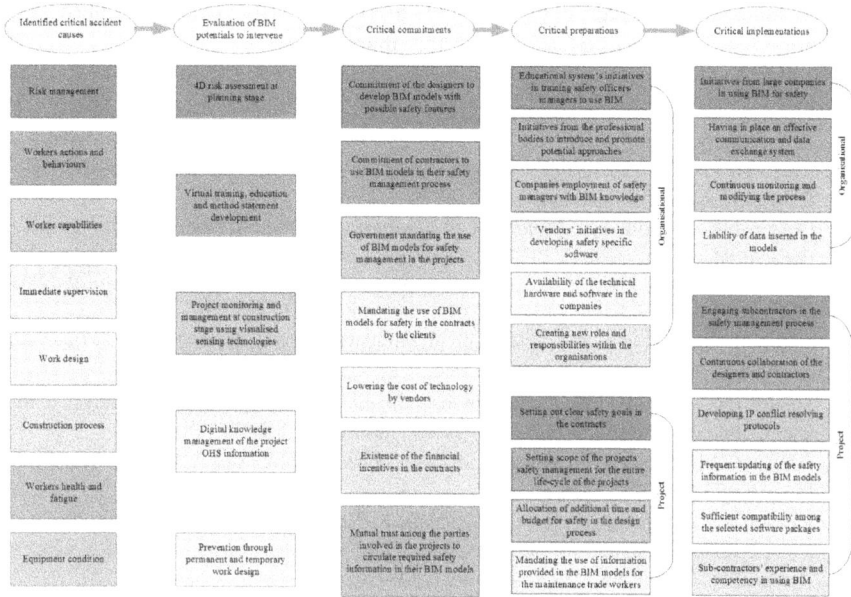

Figure 6.7 Innovation adoption model for diffusion of BIM in the WHS management of construction projects (factors are organised from the highest to lowest criticality)

The project implementation of an innovation requires that the contractor and the design team collaborate and monitor the models during the execution of the projects. These two teams should develop IP conflict-resolving protocols to facilitate the exchange of models. On the other hand, sub-contractors must be engaged in the safety management process and provided with the required information through models. Models developed for the WHS management of projects are required to be updated frequently by contractors.

6.8   Summary

Using a mixed-method research approach, this chapter examined the BIM adoption process in the Australian construction industry. A description of the study's process was provided, which included a description of how data were collected, the methodologies used to analyse data and the results that were achieved. An analysis of the existing literature revealed 27 critical factors for successful adoption. Through the semi-structured interviews as well as questionnaires, these factors were revised according to the current realities of the industry. Based on these findings, a number of parties involved in the process are required to take action, including clients, governments, contractors, design teams, procurement methods, technology providers and educational institutions. Furthermore, the role that each of these parties is required to take is described and discussed. By building off the findings from Chapter 5 regarding BIM applications for construction safety and the findings of this chapter, this research book describes a comprehensive model of innovation adoption for diffusion of BIM within safety management.

References

1. Creswell JW. *A Concise Introduction to Mixed Methods Research*. Thousand Oaks, California: Sage Publications; 2014.
2. Denzin NK, Lincoln YS. *Handbook of Qualitative Research*. London: Sage Publications; 1994.
3. Morse JM. *Critical Issues in Qualitative Research Methods*. London: Sage Publications; 1994.
4. Giddings LS. Mixed-methods research: positivism dressed in drag? *Journal of Research in Nursing*. 2006;11(3):195–203.
5. Shields PM, Tajalli H. Intermediate theory: the missing link in successful student scholarship. *Journal of Public Affairs Education*. 2006;12(3):313–34.
6. Petticrew M, Roberts H. *Why Do We Need Systematic Reviews?* Oxford: Blackwell Publishing Ltd; 2008: 1–26 p.
7. Pawson R, Greenhalgh T, Harvey G, Walshe K. Realist review – a new method of systematic review designed for complex policy interventions. *Journal of Health Services Research & Policy*. 2005;10(1_suppl):21–34.
8. Chong H-Y, Lee C-Y, Wang X. A mixed review of the adoption of building information modelling (BIM) for sustainability. *Journal of Cleaner Production*. 2017;142:4114–26.
9. Webster J, Watson RT. Analyzing the past to prepare for the future: writing a literature review. *MIS Quarterly*. 2002;26(2):xiii–xxiii.
10. Zou Y, Kiviniemi A, Jones SW. A review of risk management through BIM and BIM-related technologies. *Safety Science*. 2017;97:88–98.
11. Eastman C, Lee J-M, Jeong Y-S, Lee J-K. Automatic rule-based checking of building designs. *Automation in Construction*. 2009;18(8):1011–33.
12. Vakilinezhad M, Dias P, Ergan S, editors. Achieving model-based safety at construction sites: BIM and safety requirements representation. *Proc of the 33rd CIB W78 Conference*. Brisbane, Australia; 2016.
13. Zhou W, Heesom D, Feng A. An interactive approach to collaborative 4D construction planning. *Journal of Information Technology in Construction (ITCon)*. 2010;14(5):30–47.
14. Golizadeh H, Hon CKH, Drogemuller R, Hosseini MR. Digital engineering potential in addressing causes of construction accidents. *Automation in Construction*. 2018;95:284–95.
15. Albert A, Hallowell MR, Kleiner B, Chen A, Golparvar-Fard M. Enhancing construction hazard recognition with high-fidelity augmented virtuality. *Journal of Construction Engineering and Management*. 2014;140(7):04014024.
16. Fang Y, Teizer J, Marks E, editors. A framework for developing an as-built virtual environment to advance training of crane operators. *Construction Research Congress 2014*. Construction in a Global Network; 2014.
17. Guo HL, Li H, Li V. VP-based safety management in large-scale construction projects: a conceptual framework. *Automation in Construction*. 2013;34:16–24.
18. Park C-S, Kim H-J. A framework for construction safety management and visualization system. *Automation in Construction*. 2013;33:95–103.

19. Zou Y, Kiviniemi A, Jones SW. Developing a tailored RBS linking to BIM for risk management of bridge projects. *Engineering, Construction and Architectural Management.* 2016;23(6):727–50.

20. Ivory C, Alderman N. Can project management learn anything from studies of failure in complex systems? *Project Management Journal.* 2005;36(3):5–16.

21. Gambatese JA, Hallowell M. Enabling and measuring innovation in the construction industry. *Construction Management and Economics.* 2011;29(6):553–67.

22. Yang Y, Li B, Yao R. A method of identifying and weighting indicators of energy efficiency assessment in Chinese residential buildings. *Energy Policy.* 2010;38(12):7687–97.

23. Yoon KP, Hwang C-L. *Multiple Attribute Decision Making: An Introduction.* Thousand Oaks, California: Sage Publications; 1995.

24. Lee S, Park G, Yoon B, Park J. Open innovation in SMEs – an intermediated network model. *Research Policy.* 2010;39(2):290–300.

25. Hosseini M, Banihashemi S, Chileshe N, Namzadi MO, Udaeja C, Rameezdeen R, et al. BIM adoption within Australian Small and Medium-sized Enterprises (SMEs): an innovation diffusion model. *Construction Economics and Building.* 2016;16(3):71.

26. Gupta B, Dasgupta S, Gupta A. Adoption of ICT in a government organization in a developing country: an empirical study. *The Journal of Strategic Information Systems.* 2008;17(2):140–54.

27. Aksorn T, Hadikusumo BH. Critical success factors influencing safety program performance in Thai construction projects. *Safety Science.* 2008;46(4):709–27.

28. Liu H, Skibniewski MJ, Wang M. Identification and hierarchical structure of critical success factors for innovation in construction projects: Chinese perspective. *Journal of Civil Engineering and Management.* 2016;22(3):401–16.

29. Banihashemi S, Hosseini MR, Golizadeh H, Sankaran S. Critical success factors (CSFs) for integration of sustainability into construction project management practices in developing countries. *International Journal of Project Management.* 2017;35(6):1103–19.

30. Tymvios N, Gambatese J, Sillars D, editors. Designer, contractor, and owner views on the topic of design for construction worker safety. *Construction Research Congress 2012: Construction Challenges in a Flat World.* West Lafayette, IN; 21–23 May 2012: 341–55.

31. Fang D, Chen Y, Wong L. Safety climate in construction industry: a case study in Hong Kong. *Journal of Construction Engineering and Management.* 2006;132(6):573–84.

32. Robson C. *Real World Research* (2nd ed.). Malden: Blackwell Publishing; 2002.

33. Seidman I. *Interviewing as Qualitative Research: A Guide for Researchers in Education and the Social Sciences.* New York: Teachers College Press; 2013.

34. Bolger F, Wright G. Improving the Delphi process: lessons from social psychological research. *Technological Forecasting and Social Change.* 2011;78(9):1500–13.

35. Guest G, Bunce A, Johnson L. How many interviews are enough? An experiment with data saturation and variability. *Field Methods.* 2006;18(1):59–82.

36. Francis JJ, Johnston M, Robertson C, Glidewell L, Entwistle V, Eccles MP, et al. What is an adequate sample size? Operationalising data saturation for theory-based interview studies. *Psychology and Health.* 2010;25(10):1229–45.

37. Bethlehem J. *Applied Survey Methods: A Statistical Perspective.* Hoboken, New Jersey: John Wiley & Sons; 2009.

38. Ho R. *Handbook of Univariate and Multivariate Data Analysis and Interpretation with SPSS.* New York: Chapman and Hall/CRC; 2006.

39. Hair Jr JF, Hult GTM, Ringle C, Sarstedt M. *A Primer on Partial Least Squares Structural Equation Modeling (PLS-SEM).* Thousand Oaks, California: Sage Publications; 2016.

40. Xiong B, Skitmore M, Xia B. A critical review of structural equation ccident applications in construction research. *Automation in Construction*. 2015;49:59–70.
41. Slaughter ES. Implementation of construction innovations. *Building Research & Information*. 2000;28(1):2–17.
42. Barlow J. Innovation and learning in complex offshore construction projects. *Research Policy*. 2000;29(7–8):973–89.
43. Tam CM, Zeng SX, Deng ZM. Identifying elements of poor construction safety management in China. *Safety Science*. 2004;42(7):569–86.
44. Blayse AM, Manley K. Key influences on construction innovation. *Construction Innovation*. 2004;4(3):143–54.
45. Ganah A, John GA. Achieving level 2 BIM by 2016 in the UK. *Computing in Civil and Building Engineering*. 2014;2014:143–50.
46. BSI: PAS 1192–6. *Specification for Collaborative Sharing and Use of Structured Health and Safety Information using BIM*. The British Standards Institution (BSI); 2018.
47. Gann D. Putting academic ideas into practice: technological progress and the absorptive capacity of construction organizations. *Construction Management & Economics*. 2001;19(3):321–30.
48. Winch G. Zephyrs of creative destruction: understanding the management of innovation in construction. *Building Research & Information*. 1998;26(5):268–79.
49. Nam CH, Tatum CB. Leaders and champions for construction innovation. *Construction Management & Economics*. 1997;15(3):259–70.
50. Gann DM, Wang Y, Hawkins R. Do regulations encourage innovation? The case of energy efficiency in housing. *Building Research & Information*. 1998;26(5):280–96.
51. Gambatese JA, Behm M, Hinze JW. Viability of designing for construction worker safety. *Journal of Construction Engineering and Management*. 2005;131(9):1029–36.
52. Gu N, London K. Understanding and facilitating BIM adoption in the AEC industry. *Automation in Construction*. 2010;19(8):988–99.
53. Schulte PA, Rinehart R, Okun A, Geraci CL, Heidel DS. National prevention through design (PtD) initiative. *Journal of Safety Research*. 2008;39(2):115–21.
54. Tymvios N, Gambatese JA. Perceptions about design for construction worker safety: viewpoints from contractors, designers, and university facility owners. *Journal of Construction Engineering and Management*. 2015;142(2):04015078.
55. Arslan M, Riaz Z, Kiani AK, Azhar S. Real-time environmental monitoring, visualization and notification system for construction H&S management. *Journal of Information Technology in Construction*. 2014;19:72–91.
56. Bahn S. Workplace hazard identification and management: the case of an underground mining operation. *Safety Science*. 2013;57:129–37.
57. Benjaoran V, Bhokha S. An integrated safety management with construction management using 4D CAD model. *Safety Science*. 2010;48(3):395–403.
58. Choe S, Leite F. Temporal and spatial information integration for construction safety planning. *Journal of Computing in Civil Engineering*. 2015:483–90.
59. Dong C, Wang F, Li H, Ding L, Luo H. Knowledge dynamics-integrated map as a blueprint for system development: applications to safety risk management in Wuhan metro project. *Automation in Construction*. 2018;93:112–22.
60. Choi B, Lee HS, Park M, Cho YK, Kim H. Framework for work-space planning using four-dimensional BIM in construction projects. *Journal of Construction Engineering and Management*. 2014;140(9).
61. Hasan S, Zaman H, Han S, Al-Hussein M, Su Y, editors. Integrated building information model to identify possible crane instability caused by strong winds. *Construction Research Congress 2012*. Construction Challenges in a Flat World; 2012.

62. Huang C, Wong CK. Optimisation of site layout planning for multiple construction stages with safety considerations and requirements. *Automation in Construction.* 2015;53:58–68.

63. Hardie M, Newell G. Factors influencing technical innovation in construction SMEs: an Australian perspective. *Engineering, Construction and Architectural Management.* 2011;18(6):618–36.

7 BIM for the Future of Construction WHS

7.1 Review of Processes, Objectives and Findings

7.1.1 Research Background, Aim and Methods

The potential uses of BIM can drastically alter WHS practices in the construction industry, which is currently suffering a sad reputation due to its high number of associated injuries and fatalities. BIM facilitates project information exchange and management and supports better collaboration and project planning by enabling virtual visualisation of the construction process (1, 2). All of these attributes have resulted in an exponential growth in interest towards the digitalised management of construction safety in the past five years (3).

BIM-enabled approaches towards WHS management are extensive and, at the same time, relatively new to the construction industry of Australia. Therefore, the diffusion of such innovative interventions with the current practice of the industry in a practical manner requires the proper identification of effective areas and evaluation of their impact on key criteria of the projects and organisations. In the construction industry context, Slaughter (4) describes innovation as the actual use of a nontrivial alteration in terms of an enhancement of a system or working procedure that is new to the corresponding organisation. As suggested by Slaughter (4), the first step for implementation of an innovation in a project or organisation is to identify the areas that require intervention. Therefore, in the case of the diffusion of BIM-enabled approaches to reduce fatalities in the construction sector, unearthing how construction causalities occur is key to prioritising areas that require intervention (5). Furthermore, BIM is not a magic bullet that can hit all WHS targets of construction fatalities. Such adoption in the industry requires a careful evaluation of the potential of BIM to be effective in terms of the identified WHS concerns. Finally, the adoption of innovations in construction organisations and projects requires an understanding of the CSFs of commitments, preparations and implementation methods (6). These factors are fundamental for tackling the major obstacles of leveraging BIM applications towards improving the WHS performance of construction projects.

This research book was aimed to develop an innovation adoption model for the diffusion of BIM in the WHS management of the construction industry to reduce occupational fatalities. To achieve this aim, the following objectives were set:

DOI: 10.1201/9781003224853-7

1) To identify the causes of fatal accidents in the construction industry.
2) To evaluate the effectiveness of BIM-enabled approaches to reduce the identified causes of fatal accidents in the construction industry.
3) To determine the CSFs of adopting BIM to improve construction WHS performance.
4) To develop an innovation adoption model to integrate BIM's WHS aspects with the current practice of the industry.

In order to achieve the designed aim and objectives of this research book, a mixed-method approach was designed for each objective. For the first objective of the research, a $QUAL{\rightarrow}QUAN$ approach was undertaken. Having access to the NCIS data made it possible to identify the fatal accident causations of real accident cases by conducting a thematic content analysis. To identify the most critical causations, standardised degree centralities were then calculated for each of the accidents, and this led to the identification of medium- and high-priority central factors for the categories of accident mechanisms. For Objectives 2, 3 and 4 of the study, another $QUAL{\rightarrow}QUAN$ approach was considered. First, a preliminary review of the literature was conducted to identify BIM's potential for controlling accident causations and determine CSFs for BIM adoption in construction WHS management. The findings of the literature review were then subjected to a qualitative analysis approach through interview sessions with industry experts from Australia. To validate the outcomes of the qualitative approach, the quantitative approach of an expert survey was carried out. The interview results were subjected to thematic content analysis and the collected survey results were subjected to PLS-SEM analysis.

7.1.2 Review of Research Processes and Findings

As mentioned, the overarching aim of this research was to develop an innovation adoption model for diffusion of BIM in the WHS management of the construction industry and reduce occupational fatalities. Hence, the key processes and findings presented in Chapters 4 and 5, which enabled achievement of the research aim, are discussed in the following sub-sections.

7.1.2.1 Objective 1: To Identify the Causes of Fatal Accidents in the Construction Industry

The primary focus of this objective was to understand the nature of fatal construction accidents and to identify the areas that require the intervention of an innovation. The Australian construction industry was used as a case study for this research. National accident surveillance reports published by organisations such as Safe Work Australia do not provide enough detailed information for researchers to develop possible preventive measures. For instance, Safe Work Australia (7) reports on construction accidents classify fatal accident causations by their mechanisms and occupation of the deceased workers and lack analysis of the causations behind the accidents.

Reviewing the existing literature on accident theories showed that there has been a paradigm shift among safety researchers from examining single causations of accidents to systematic accident models, as these are more robust and comprehensive. Unlike sequential models, systemic accident causation models describe the existence of dynamic interaction among cultural and organisational factors in creating a hazardous situation (8). For construction accidents, the ConAC model developed by a research team from the Loughborough University was found to be the only well-known systematic accident model that had been employed by several researchers to diagnose construction accident causations (1, 8–10). Therefore, the current study took advantage of the ConAC model to identify the critical areas in the occupational fatalities of the Australian construction industry.

Out of 287 fatal accident cases collected from the NCIS database for the period of 2007–2016, 105 cases were deemed suitable for the thematic content analysis, as the removed cases did not contain coroner findings regarding the accidents and police reports did not describe the causations behind the accidents. Accident cases were categorised and described based on their locations, times and accident mechanisms. Only two cases from NSW were suitable for the analysis, and most of the accident cases were from the period of 2007–2014. Fall from height and contact with electricity were the two major mechanisms of accidents among the collected cases, as the Safe Work Australia (7) report also indicated.

Using the terminologies of the ConAC model described by Behm (11) and later by Golizadeh, Hon (1), the fatal accident cases were set for deductive content analysis. The analysis found that the highest proportion of causations was related to immediate supervision (in 60 cases) and workers' actions and behaviour (in 58 cases), followed by risk management (in 54 cases), construction processes (in 48 cases) and permanent work design (in 45 cases) from the originating influences group. In addition, causations were analysed for accident mechanisms. For instance, in fall-from-height accidents, immediate supervision, permanent work design, workers' actions and behaviours and construction process were the highest contributing factors. In contact-with-electricity accidents, permanent work design, construction process and risk management were found as frequent factors in the cases.

Many past studies have considered accidents as the result of a network of causations, whereby their linkage leads to an accident (8–10, 12). Therefore, those accidents that had the highest number of linkages were considered the most critical causations, and degree centrality was deemed to be an appropriate approach to discover such causations (13, 14). In order to compare the value of centralities, standardised degree centrality values for the causations of each accident network were calculated. Standardised degree centrality values were then categorised for each accident mechanism in a four-scale format of very low, low, medium and high centrality. Based on the findings, risk management, workers' actions and behaviours, workers' capabilities, immediate supervision, temporary and permanent work design, construction process, workers' health and fatigue and equipment conditions were the most repeated high and medium central causations in the accident mechanisms. These causations were considered critical areas that required intervention via an innovation, which in the case of this study, is employing BIM.

7.1.2.2   Objective 2: To Evaluate the Effectiveness of BIM-Enabled Approaches and Reduce the Identified Causes of Fatal Accidents in the Construction Industry

The second stage in the innovation adoption theory was to evaluate innovation capabilities to elevate the identified areas for intervention. Hence, BIM capabilities for construction WHS management were identified by reviewing the existing peer reviewed journal papers on the topic. Six types of BIM applications that may support construction WHS management were found: 3D tools for preliminary risk assessment; automatic/semi-automatic model checkers; 4D (3D+time) construction planning; knowledge management systems; AR and VR; and RTLS.

The next step for this objective was to identify applications of BIM that could specifically improve the identified critical accident causations. Semi-structured interviews with highly experienced industry experts were employed as an appropriate approach to collect sufficient data. In essence, this study used purposeful sampling techniques for the selection of participants instead of random sampling (15). The population of interest for this part of the study consisted of construction safety and project managers from the contractor sector, as well as BIM managers, architects and design engineers from the consultant party that had engaged with several BIM-based projects and were familiar with the regulations and industrial settings of the Australian construction industry. Having at least 10 years' experience was identified as the main criterion for the selection of participants, along with their availability in the planned interview period. Overall, six highly experienced industry experts were interviewed and provided with sufficient information regarding the progress of the research and existing approaches found in the literature. This was to ensure that the study was consistent in structure and data collection processes and therefore contributed to the reliability of collected data (16).

After conducting each semi-structured interview, the recorded interviews were transcribed and subjected to thematic content analysis to identify BIM applications related to the critical accident causations. The interviews continued until no new codes emerged in two consequent sessions. After finalising the results, five main domains were recognised that may be useful in improving the critical accident causations. The first approach was virtual training, education and method statement development, including:

- training workers in an interactive virtual environment that can involve all existing risks rather than the traditional classroom-based settings;
- testing and improving worker safety culture in situations such as reporting unsafe conditions on the site, recognising the required PPE, communicating safety concerns with their colleagues, etc.; and
- developing safe work method statements for companies and trade unions.

The second approach was prevention through permanent and temporary work designs that include:

- automated or semi-automated checking of the building designs in BIM models to avoid the unsafe designs or identifying possible hazards as early as possible; and

- automated or semi-automated designing and checking the design of the on-site temporary structures using validated algorithms to avoid design related failures.

The third approach was 4D risk assessment at the planning stage, including:

- suitability assessment of the dynamic equipment/machinery using 4D BIM simulations;
- planning ahead for the required materials and equipment for safer execution of the projects using 4D BIM planning; and
- assessing workers and sub-contractors' capabilities by conducting a walk-through of the 4D simulations and discussing their approach on safely performing their tasks.

The fourth approach dealt with project monitoring and management at the construction stage using visualisable sensing technologies, including:

- monitoring workers and equipment through attached location tracking sensors that are connected to a BIM model and controlled by the project management team;
- performing frequent monitoring of the site and equipment conditions using flying drones;
- communicating safety issues through software such as BIM 360 Field, which circulates information through tablets and smartphones and informs the responsible parties to take action;
- monitoring workers' health and fatigue conditions by using remote health monitoring sensors that can capture the location and health condition of the workers and show them in a BIM model; and
- monitoring environmental conditions such as temperature, humidity and wind speed in indoor and outdoor locations through sensors linked to a BIM model.

The last approach was digital knowledge management of the project's WHS information, including:

- Using BIM as a digital library of the site's condition. An external database that stores site safety information is created, and the required safety information is linked to the objects in the BIM model. This provides project/safety managers with a comprehensive understanding of the situation.
- Using BIM as a digital library of equipment/machinery maintenance information. An external database is created to store operation and maintenance information about the equipment/machinery used in the 4D BIM models.
- Using BIM as a digital library of the workers' safety track record. An external database of the workers' health and performance track records is created and linked to the 4D BIM models. This shows workers' health and safety performance records when assigned to different activities.
- Using BIM models for suppliers to assess the suitability of the materials and equipment they provide for construction sites.

To evaluate the effectiveness of the identified approaches/types of BIM on critical causations, a survey was conducted among a larger subset of industry experts. Using PLS-SEM analysis, BIM applications under 4D risk assessment were found to be the most effective at the planning stage. Accordingly, applications related to virtual training, education, method statement development, project monitoring and management at the construction stage using visualised sensing technologies, digital knowledge management of the project WHS information and prevention through permanent and temporary work design approaches were, in that order, the most to the least effective at the current stage.

7.1.2.3   *Objective 3: To Determine the CSFs for Adopting BIM to Improve Construction WHS Performance*

This objective dealt with identifying CSFs at the commitment, preparation and implementation stages of the innovation adoption model devised in this study. The research method conducted for this objective was the same as that used for Objective 2. First, a comprehensive review of the existing literature was conducted on the CSFs for successful implementation of BIM for construction WHS management. Since there were limited sources due to the newness of the topic, a search was undertaken to find literature related to CSFs for the adoption of ICT, BIM and innovations in construction WHS management. The search yielded 27 CSFs related to different stakeholders of the contractor, design team, client, government, professional and educational bodies and technology providers.

As for Objective 2, semi-structured interviews were conducted to identify CSFs that were specifically related to the adoption of BIM to reduce the incidence of fatalities in the construction industry of Australia. Using the same research method, the CSFs were identified by reaching saturation in data. Regarding the commitment stage, the following seven CSFs were required:

- commitment of designers to develop BIM models with possible safety features;
- mutual trust among the parties involved in the projects to circulate the required safety information in their BIM models;
- workers' agreement regarding the storage and use of their health and safety records;
- commitment of contractors to use BIM models in their safety management system;
- reduction in the cost of technology by vendors to make the technology usable for small and medium-sized companies;
- Australian government mandating for the use of BIM models for safety management of the projects;
- mandating for the use of BIM models for safety in the contracts by the clients; and
- the existence of financial incentives in the contracts for using BIM in the safety management process.

In the preparation stage, two categories of preparations were identified. The six types of identified organisational preparations were:

- TAFE and university initiatives in training safety officers/managers to use BIM;
- initiatives from the Master Builders Association for introducing possible approaches to the companies;
- vendors' initiatives in developing software that can identify unsafe designs according to national safety regulations;
- availability of technical hardware and software in the companies;
- creation of new roles and responsibilities within organisations for facilitating the implementation of BIM for safety management; and
- employment of safety managers with BIM knowledge in the companies.

Project-related preparations included:

- setting clear safety goals in the contracts for using BIM in the safety management process;
- setting the scope of the project's safety management for the entire life cycle of the project from the first draft;
- allocating additional time and budgeting for safety in the design process;
- mandating the use of information provided in the BIM models for maintenance trade workers.

At the last step of the innovation adoption model of the research, two types of implementation are required. For organisational implementation, four CSFs were identified: initiatives from large companies as the front-liners in the industry in using BIM for safety, liability of data inserted in the models for each of the parties involved, continuous monitoring and modification of the process of using BIM for safety management, and existence of an effective communication and data exchange system in place among the people involved in the safety management process of the projects.

Six CSFs were found to be required for project-wise implementations:

- sufficient compatibility among the selected software packages to transfer safety features from one to another;
- development of IP conflict-resolving protocols to facilitate the exchange of models between designers and contractors;
- continuous collaboration of the designers and contractors in monitoring and modifying the models during the execution of the projects;
- frequent updating of the safety information in the BIM models;
- sub-contractor experience and competency in using BIM; and
- sub-contractor engagement in the safety management process and access to required information through models.

Finally, the findings of the qualitative approach were set to a broad survey among industry experts. The results employing PLS-SEM to quantify the importance of the identified CSFs are shown in Figure 6.7. In the commitment stage, factors related to the commitment of designers to develop BIM models with possible safety features and the commitment of contractors to use BIM models in their safety management process had the highest importance. Significant preparations included the educational system's initiatives in training safety officers/managers to use BIM and initiatives from professional bodies to introduce and promote potential approaches of organisations while setting out clear safety goals in the contracts and setting the scope of the project's safety management for the entire life cycle of the project. The weightiest project-related and organisational implementation CSFs were engaging sub-contractors in the safety management process, promoting continuous collaboration of the designers and contractors, creating initiatives from large companies in using BIM for safety, and having an effective communication and data exchange system in place.

7.1.2.4 Objective 4: To Develop an Innovation Adoption Model to Integrate BIM's WHS Aspects with the Current Practice of the Industry

The overarching objective of this research was to develop a framework that describes the critical stages for the adoption of BIM to reduce fatalities in the current Australian construction industry. Because practice and research may evolve over years, the identified CSFs and BIM applications were considered recent. Combining the findings of Objectives 1, 2 and 3 of the research into one framework resulted in the development of the innovation adoption model of this research (see Section 5.4 and Figure 5.2). Objective 1 fulfilled the first step of the innovation adoption model by identifying the areas that are currently the most critical for fatal occupational accidents in Australian construction projects. This objective involved analysing real fatal accident cases that were investigated by coroners. Objective 2 focused on how BIM can improve those critical areas by introducing the potentially effective applications of BIM. Combining the recent advances of BIM in construction WHS management and experts' opinions led to the identification of BIM applications at the current maturity level that could improve the identified critical WHS areas. Objective 3 of the research formed the remaining three stages of the innovation adoption model, which were commitment, preparation and implementation. Due to the novelty of the topic, CSFs from similar topics were collected from the literature and modified by industry experts to be suitable for the topic of this research. As a result, a five-stage framework was developed that suits Australian construction projects and organisations in reducing the number of occupational fatalities.

7.2 Originality, Contributions, Implications and Limitations

7.2.1 Originality

Contributions to the body of knowledge on this in three originality measures are described as follows.

- *Conducting research in an unexplored area.* Outcomes of Objective 1 of the research present causations behind recent occupational fatalities in Australia that were previously investigated by Cooke and Lingard (17) in a similar method for accidents took place between 2000 to 2009, and there were no recent investigations on this area. This allowed for understanding of the current occupational accident causations in this study and will support future research in developing preventative measures. Objective 2 of the research included a search for BIM-enabled applications within peer-reviewed journals, and these were categorised by their potential for preventing construction accident causations (1). There was no previous research on this area, and this study therefore provides better understanding of current BIM applications for construction WHS management and the gaps remaining for future investigations. In addition, Objective 2 assessed the effectiveness of BIM applications at the current stage, which had not previously been investigated. Guo, Yu (3) noted that the popularity of BIM-enabled WHS management research has exponentially increased since 2012, and yet no research has assessed the effectiveness of these applications for controlling accidents. This enables researchers to direct their focus to the areas where BIM can be practical and avoid unnecessary areas. Overall, there is a very limited number of innovation adoption models for construction WHS management, and no previous research has been undertaken regarding the adoption of BIM to improve construction WHS management. This is an important research gap, as many countries are launching their BIM standards for construction WHS management, and yet a successful adoption process remains unexplored. Objectives 3 and 4 of the research tackled this lack of information by exploring CSFs for successful diffusion of BIM in construction projects and organisations to reduce occupational fatalities. The framework established in Objective 4 is the first of its kind in Australia and describes the adoption process in five steps: Identification of critical causations, evaluation of BIM capabilities to control the critical causations, commitment CSFs, preparations CSFs and implementation CSFs.

• *Showing a novel empirical approach.* Handfield and Melnyk (18) noted that knowledge creation is generally handled by the emergence of a new theory or by partial or whole rebuttal and/or modification of an existing theory through the employment of empirical data. This research modified Slaughter's (19) innovation adoption theory to achieve the overarching aim, which was developing a BIM adoption model for the construction industry of Australia to reduce occupational fatalities. In addition, diffusion of a well-established accident theory with the innovation adoption theory was another theoretical contribution of this research that supported the identification stage of the model. This allowed for the identification of the most central accident causations among fatal occupational accidents in the Australian construction sector. Previous studies, such as those by Behm and Schneller (9), Gibb, Lingard (8) and Cooke and Lingard (17) used the ConAC terminologies to determine the causations behind accidents in different regions while neglecting the systematic connections of the causations. As many systematic accident models, such as those of Reason (20) and Gibb, Haslam (12) suggest, eliminating those accidents that have more connections can significantly reduce the likelihood of accidents.

• *Genuine synthesis*. This research yielded a set of empirical findings that are novel. First, this research took advantage of the ConAC model's connections among accident causations to develop an accident network of the analysed cases. This led to identifying critical areas that require intervention, whether using BIM or other controlling measures to reduce the number of occupational fatalities. In addition, previous studies into construction accident causations in Australia have analysed the accident cases for the period of 2000–2009 (17), and yet there was no recent practice in this period. Objectives 2, 3 and 4 involved syntheses of the literature survey and viewpoints of industry experts regarding the effectiveness of BIM applications to mitigate the critical accident causations and CSFs for the adoption of BIM and reduce fatalities in Australian construction projects. Although there have been many studies on different applications of BIM for construction WHS management, no research has assessed BIM applications or CSFs for diffusing this innovation in the industry. Hence, the genuine synthesis rule for research was satisfied, which Walker (21) described as "making a new interpretation of existing material" (p. 150). Last, but not least, the BIM adoption model of this research represents the first attempt in this area, and no previous studies have described the process of such adoption within the construction industry of Australia or any other countries.

7.2.2   Implications for Practice

Implications for practice are another expectation for research to describe the practical knowledge extracted from this research. Yi and Chan (2013) noted that such contributions are of importance in the construction management of the built environment because there is an urgent necessity for elevating its efficiency. The implications of the current research for practice are described herewith according to taxonomies described by Bartunek and Rynes (22), which include enhanced awareness, potential audience and identification of new learning areas.

• *Potential audience*. Drawing upon the practical applications of the BIM adoption framework to reduce occupational fatalities, contractor companies can be considered the first-tier audience, as they hold most of the liability in accidents. Contractors are required to pay further attention to the critical WHS areas that lead to fatal accidents, be aware of the potential applications of BIM to reduce such fatalities and address the CSFs required for a successful adoption of BIM in their organisation and projects. Second tier audiences are the government and local organisations that are seeking to employ BIM as an innovative approach to improve the WHS management of the construction industry. The recent introduction of PAS 1192–6 in the UK (23) may encourage the local and federal governments of other countries to develop an adoption framework for employing BIM for construction WHS management. This research also can draw the attention of US and UK practitioners, as the rate of occupational fatalities in the construction industries of these countries is higher than that of all other industries.

- *Enhanced awareness.* Due to growing concerns about the number of occupational fatalities in the construction industry, this research identified the critical areas that require improvement. Risk management, workers' actions and behaviours, worker capabilities, immediate supervision, work design, construction process, workers' health and fatigue and equipment conditions were found to be the most critical areas that the industry is required to be aware of. Additionally, this research raises awareness among the parties involved in the project that collaboration between several parties is required to achieve a smaller number of fatalities. The model developed in this study shows that without the support of other parties, contractors do not have the ability to handle the adoption process. This study also highlights the rule of technology providers, as the current applications of BIM do not suit the requirements of safer practice. This research has pinpointed the areas that technology providers should pay more attention to and areas that are executed safely at the current stage.
- *New learning areas identification.* Underlining the practical problems and results of research is a resource of learning and gaining further knowledge for industry audiences. Revealing the critical causations behind occupational fatalities can challenge the industry audience to pay further attention to those areas. The identified CSFs are also challenges that the Australian construction industry faces in the adoption process of BIM for WHS management. The current research can also be informative for practitioners of other countries, as many countries employ BIM in their projects, and yet construction stands at the top of the list of hazardous industries.

7.2.3 Limitations

This section acknowledges the limitations of the current research despite the important contributions it makes to the current body of knowledge. Therefore, the limitations of this research are as follows:

- First and foremost, the current study can only be considered valid in the context of the Australian construction industry. The BIM applications identified in the current research were discussed and devised by industry experts to tackle the accident causations found for occupational fatalities of Australia. As many previous studies have shown, WHS management issues may differ from one country to another (8). The critical commitments, preparations and implementation factors were also modified for the context of current Australian construction practice. These factors may vary from time to time, as technology and working environments are dynamic.
- Accidents covered in this research are fatal occupational accidents, and the findings may or may not apply to non-occupational or non-fatal occupational accidents. Due to the availability of data for fatal occupational accidents and the limited duration and resources of the study, the scope of this research was set to this type of accidents. However, the procedures used for this research can be reused to develop innovation adoption models to reduce other types of accidents.

- The accident cases selected for the current research did not involve most cases in NSW. This was due to the lack of coroner's reports among the NSW cases, as this was one of the case selection criteria for this research. Overall, 182 accident cases found for the selected duration of the research were deleted due to this criterion; however, the trends in accident mechanisms found among the selected cases was similar to the national trends.
- Although the first objective of the study was satisfied by analysis of real accident cases, the rest of the objectives were addressed through opinion-based approaches of semi-structured interviews and a questionnaire survey. This might be a drawback of the current research method when compared with conducting action-based research. This research could be used as a base for future action-based research in the Australian construction industry.

The majority of the participants who completed the questionnaire survey had less than five years of experience using BIM in their usual practice, even though they had many years of industry experience. This is probably due to construction organisations in Australia having only recently adopted BIM.

7.3   Knowledge Gaps Within the Domain of BIM and WHS

The limitations found in this research can pave the way for future research in this area. Zou and Sunindijo (24) noted that research has a cyclical feature and does not stop at a point. Thus, the recommendations for future research driven by this study are described as follows:

- The BIM adoption model devised in this study deals with occupational fatalities in the Australian construction industry. The procedure adopted for the current research could be readopted for a wider purpose of BIM and digital engineering adoption to improve national WHS management. Additionally, this procedure could be good practice for international adoption processes, as many countries, such the UK, are initiating BIM implementation for construction WHS management (23). It is noteworthy that for the successful implementation of any innovation, appropriate diagnosis of the problems is crucial, and this may vary from one country to another.
- The theoretical approach of the current study provides a sound structure for future research into WHS management. Integrating systematic accident models with innovation adoption theories can potentially support a sound background for future research and lead to the modification of adoption frameworks. Accident models such as the ConAC system not only provide the terminology for comprehensive accident analysis but also describe the importance of the linkages among the actors. The latter requires further attention from researchers, as accidents take place through a chain of events and the central actors are the most critical ones. Thus, measuring the degree centralities of accident causations is a good approach to identifying central links.
- As mentioned by Guo, Yu (3) and Golizadeh, Hon (1), research into the domain of BIM-enabled WHS management has grown over the last five years.

However, many of these studies have overlooked what the current practice of the industry requires most and which areas of WHS management are currently doing well. Identification of the critical WHS management areas in this study led to recognition of what BIM-enabled research should be focused on in the future. Thus, further research is required on the topics of 4D risk assessment at the planning stage, virtual training, education and method statement development, project monitoring and management at the construction stage using visualised sensing technologies, digital knowledge management of the project WHS information and prevention through permanent and temporary work design.

- CSFs identified in the current research can be tested in real-world practice and evaluated. Identifying these CSFs did not benefit much from the previous research due to limited resources and potential future research could investigate CSFs in different regional contexts other than Australia.
- Finally, the innovation adoption model of the research was mostly developed by employing opinion-based approaches, and as such, conducting field research, such as action research, could further enhance the accuracy of the model.

7.4   Summary

This chapter summarised the existing gaps in the literature, research aims and objectives and the process undertaken to address the research objectives. The findings and discussions made around the objectives of the research were also provided. Finally, this chapter outlined the contributions of this research to the body of knowledge, as well as discussing the implications for practice, recommendations for future research and limitations of the current research.

References

1. Golizadeh H, Hon CKH, Drogemuller R, Hosseini MR. Digital engineering potential in addressing causes of construction accidents. *Automation in Construction.* 2018;95:284–95.
2. Zou Y, Kiviniemi A, Jones SW. A review of risk management through BIM and BIM-related technologies. *Safety Science.* 2017;97:88–98.
3. Guo H, Yu Y, Skitmore M. Visualization technology-based construction safety management: a review. *Automation in Construction.* 2017;73:135–44.
4. Slaughter ES. Models of construction innovation. *Journal of Construction Engineering and Management.* 1998;124(3):226–31.
5. Swuste P. "You will only see it, if you understand it" or occupational risk prevention from a management perspective. *Human Factors and Ergonomics in Manufacturing & Service Industries.* 2008;18(4):438–53.
6. Banihashemi S, Hosseini MR, Golizadeh H, Sankaran S. Critical success factors (CSFs) for integration of sustainability into construction project management practices in developing countries. *International Journal of Project Management.* 2017;35(6):1103–19.
7. Safe Work Australia. *Work-Related Injuries and Fatalities in Construction, Australia, 2003 to 2013.* Australia; 2015.

8. Gibb A, Lingard H, Behm M, Cooke T. Construction accident causality: learning from different countries and differing consequences. *Construction Management and Economics*. 2014;32(5):446–59.

9. Behm M, Schneller A. Application of the Loughborough construction accident causation model: a framework for organizational learning. *Construction Management and Economics*. 2013;31(6):580–95.

10. Lingard H, Cooke T, Gharaie E. A case study analysis of fatal incidents involving excavators in the Australian construction industry. *Engineering, Construction and Architectural Management*. 2013;20(5):488–504.

11. Behm M. Relevancy of data entered into riskmaster. *NCDOT Research Project 2009–10*. East Carolina University; 2009. Available from: https://connect.ncdot.gov/projects/research/RNAProjDocs/2009-10FinalReport.pdf.

12. Gibb AG, Haslam R, Gyi DE, Hide S, Duff R. What causes accidents? *Proceedings of the Institution of Civil Engineers*. 2006;159(6):46–50.

13. Alsamadani R, Hallowell M, Javernick-Will AN. Measuring and modelling safety communication in small work crews in the US using social network analysis. *Construction Management and Economics*. 2013;31(6):568–79.

14. Pryke SD. Towards a social network theory of project governance. *Construction Management and Economics*. 2005;23(9):927–39.

15. Eisenhardt KM. Building theories from case study research. *Academy of Management Review*. 1989;14(4):532–50.

16. Saunders M, Lewis P, Thornhill A. *Research Methods for Business Students*. Essex: Pearson Education; 2009.

17. Cooke T, Lingard H, editors. A retrospective analysis of work-related deaths in the Australian construction industry. *ARCOM Twenty-seventh Annual Conference*. Association of Researchers in Construction Management (ARCOM); 2011.

18. Handfield RB, Melnyk SA. The scientific theory-building process: a primer using the case of TQM. *Journal of Operations Management*. 1998;16(4):321–39.

19. Slaughter ES. Implementation of construction innovations. *Building Research & Information*. 2000;28(1):2–17.

20. Reason J. Human error: models and management. *BMJ: British Medical Journal*. 2000;320(7237):768–70.

21. Walker DH. Choosing an appropriate research methodology. *Construction Management and Economics*. 1997;15(2):149–59.

22. Bartunek JM, Rynes SL. The construction and contributions of "implications for practice": what's in them and what might they offer? *Academy of Management Learning & Education*. 2010;9(1):100–17.

23. BSI: PAS 1192–6. *Specification for Collaborative Sharing and Use of Structured Health and Safety Information Using BIM*. The British Standards Institution (BSI); 2018.

24. Zou PX, Sunindijo RY. *Strategic Safety Management in Construction and Engineering*. Oxford: John Wiley & Sons; 2015.

Index

For Product Safety Concerns and Information please contact our EU
representative GPSR@taylorandfrancis.com
Taylor & Francis Verlag GmbH, Kaufingerstraße 24, 80331 München, Germany

www.ingramcontent.com/pod-product-compliance
Lightning Source LLC
Chambersburg PA
CBHW060313220326
41598CB00027B/4315